DAY OF THE COBRA

The true story of KAL Flight 007

JEFFREY ST. JOHN

THOMAS NELSON PUBLISHERS
Nashville · Camden · New York

Published in Nashville, Tennessee, by Thomas Nelson, Inc. and distributed in Canada by Lawson Falle, Ltd., Cambridge, Ontario.

Printed in the United States of America.

Fourth printing

Library of Congress Cataloging in Publication Data

St. John, Jeffrey.
Day of the cobra.

Includes bibliographical references.
1. Korean Air Lines Incident, 1983. 2. United States—Foreign relations—Soviet Union. 3. Soviet Union—Foreign relations—United States. 4. Aeronautics and state—Soviet Union. 5. Airspace (International law) I. Title.
E183.8.S65S72 1984 909'.096454 84-4780
ISBN 0-8407-5381-0

At all times civilization has its enemies, though they are constantly changing their guise and their weapons. The great defensive art is to detect and unmask them before the damage they inflict is fatal. . . . Survival is falsehood detected in time.

Dr. Paul Johnson
Enemies of Society
1977

To the memory of the 269 victims of Korean Airlines Flight 007, to their living relatives and loved ones, and to the nation of South Korea whose ordeal throughout its history may be as well understood by those who read this work as it was by KAL 007's most famous victim, Congressman Lawrence P. McDonald (1935–1983), this book is respectfully dedicated.

Contents

Foreword

by United States Senator Jesse Helms

When the president urged me to lead the U.S. delegation to Korea for the thirtieth anniversary of the signing of the United States–Korea Mutual Defense Treaty, I was reluctant to go. To be honest about it, the week of that trip was the only time I had set aside in August 1983 to be with my grandchildren. All the rest of the month had been very busy, leaving no time for personal enjoyment.

But since South Korea has been such a firm and faithful anti-Communist ally, I agreed to go. I insisted to the White House, however, that I would not go at the taxpayers' expense. Thus my travel was by commercial airlines, paid for with private funds.

That's why I was aboard Korean Airlines flight 015 the night it left Los Angeles on August 30, 1983. Meanwhile, the ill-fated KAL 007 plane had left New York a few hours earlier and arrived at Anchorage, Alaska, about twenty minutes before ours. Both planes were on the ground at Anchorage at the same time, for the better part of an hour.

The passengers aboard both planes got off—most of them, at any rate. A few did not. Congressman Larry McDonald (D–Ga.) was asleep and stayed aboard his plane. We did not know he was there.

United States Senator Steve Symms of Idaho and Congressman Carroll Hubbard of Kentucky had elected to take the flight with me out of Los Angeles. Senators Orrin Hatch of Utah and Ed Zorinski of Nebraska came a day later. First news reports indicated that all of us were on the same plane, the one that was shot down on September 1, 1983, by the Soviets.

When I checked in with my Washington office by telephone, I learned that telegrams, telephone calls, and letters had poured in by the hundreds. I was deeply touched by the concern expressed by so many who were praying for our safety. But the fact is we knew nothing about the other plane's fate until we arrived at Seoul. Even then the reports were confusing for the first few hours.

Representatives of the U.S. Defense Department and our intelligence people briefed us constantly. Finally the shocking news came: KAL 007 had been shot down, intentionally and premeditatedly, by the Soviets. Two hundred sixty-nine innocent people, including Congressman McDonald, were dead.

It was a sad, traumatic, and emotional period, as you can imagine. Even now, I can still see two little girls, one five years old and the other three, whom I met in the airport lounge at Anchorage. Theirs was a lovely family—the father and mother were fine, attractive young people. They were one of the most delightful young families anyone could hope to see: Neil J. Grenfell, his wife Carol Ann, and his two daughters, Nolie Ann, five, and Stacy Marie, three. The Grenfells had been vacationing in Rochester, New York, and were returning to Seoul, where Mr. Grenfell was marketing development director for Eastman Kodak.

The mother was sitting reading Bible stories to those two little girls when we entered. The smallest was sitting on her mother's lap, and the five-year-old was sitting on the arm of the chair. When the mother had finished reading to the children, I went over and introduced myself. They were soon on my lap, and we were playing a little game that I play with my grandchildren.

I will never forget those two little girls, who had a right to live and love, but who will never have that right because of this criminal, brutal, premeditated, cowardly act by the Soviet Union. I will forever remember the giggles and the laughter as I played a lighthearted game with them; three-year-old Stacy giggled and put out her little arm, saying repeatedly, "Do it again!"

And I did—over and over again. Then the loudspeaker announced the impending departure of KAL Flight 007. The two children hugged my neck, kissed me on the cheek, then scampered out the door with their parents. They boarded that doomed plane, and soared upward into the darkness.

In the quiet of the night, I can still hear those precious children laughing. I can feel their little arms around my neck. I can see them waving good-bye and blowing kisses to me as they departed.

I simply cannot accept their deaths. My mind rejects any notion that a civilized country could wantonly, intentionally, premeditatedly shoot down an unarmed plane loaded with innocent men, women, and children.

How I wish I could hear that little one say once again, "Do it again!" But that option is not open.

Tennyson put it well:

> *But O for the touch of a vanish'd hand,*
> *And the sound of a voice that is still!**

*From "Break, Break, Break."

Lawrence Patton McDonald

Introduction

***"It meant they failed to see what was true, until too late, when it simply was a question of survival."*[1]**

A. L. Rowse
Appeasement: A Study in Political Decline 1933–39

Rowse, along with Winston Churchill, was a warning voice in the political wilderness of the 1930s when British Prime Minister Neville Chamberlain and the entire English establishment embarked on a policy of appeasing Nazi Germany's Adolf Hitler. United States historian Telford Taylor maintains that Munich "is a potent and historically valid symbol of the dangers of not facing up to unpleasant realities."[2]

Dr. Lawrence P. McDonald spent his entire career in the U.S. House of Representatives warning his colleagues and the country that a refusal to face the realities of Soviet designs was to repeat the dangerous follies of the past. "Reality counts for little and illusion is frequently king," he wrote of U.S. contemporary political life five months before his death aboard Korean Airlines Flight 007 on September 1, 1983, "but in the struggle for the survival of Western civilization, it will be the real world, not illusions or delusions, that will determine which way the future will go."[3]

McDonald had a passionate concern about the future, based on his extensive and knowledgeable understanding of the past. I recall the six days we spent together attending an international conference at historic Windsor Castle in England during the first week of March 1980. On one sunny but raw day the congressman, his wife Kathryn, and I spent hours walking, talking, and wandering through London. When we reached historic Waterloo Bridge we could see Big Ben and the British Parliament buildings across the Thames River.

My notebook for that trip reveals McDonald's making a series of observations that the mindset of the U.S. establishment, when it came to the Soviets, was strikingly similar to that of the British in the 1930s prior to Munich. "The problem then and now," he said as he motioned with his hand toward the Houses of Parliament across the Thames, "was and is the refusal to believe that the past has any relationship to the present and the future."[4]

Later in a visit to the House of Commons when the British Parliament was in session, he observed it was strange that so many in both British and American political life had gone through the searing experience of Munich and World War II and yet insisted that the 1930s and the 1970s–1980s were somehow different. He also spent considerable time talking about England, not only as the cradle of the Industrial Revolution but also as the mother of the United States. As a descendant of the Scots, he pointed out with understandable pride that it was they who had led the fight for religious liberty against the British Crown, which led to the establishment of Jamestown in 1607. From this rebellion against the "divine right of kings" all other freedoms and traditions of the English-speaking people had flowed like a river that had begun with a ripple of protest.

After his death at the hands of a Soviet aircraft, I recalled that London trip as the news reports on his political career characterized him solely as an archconservative Democratic member of Congress who was a leader in the "radical right wing" John Birch Society. McDonald was fond of pointing out that the word *radical* means, in its first definition, "of or relating to the root."

During his nine years in Congress the medical physician-turned-politician was forever insisting that America's roots were in the Founding Fathers (born out of the British experience), who created the United States of America. In fact he was one of the few elected officials to write and publish a thoughtful history of the origins of the U.S. Constitution and to explain how, as a document, it remains a constellation of practical moral and political principles that time and technology have not diminished.

McDonald wrote in his work published during the 1976 Bicentennial of the Declaration of Independence:

> Ever since Franklin D. Roosevelt in the mid-1930s equated the governmental principles in the Constitution with a "horse-and-buggy" society, we began to confuse principles and material goods in our thinking. The "buggy" may have been outmoded, but political principles are eternal. Neither time nor new technology alters them.[5]

Right up to the time of his tragic and violent death McDonald pressed his arguments for the defense of Western civilization. He saw that America's arsenal should comprise a strong military defense as well as a moral and spiritual one. "The wreckage of other ages and civilizations," he wrote, "is about to remind us of follies and the nature of our enemies."

> A good example is the case of ancient Israel when it ceased to be the great nation of its heritage. The prophet Jeremiah was sent by the Lord to give warning to a people intoxicated by affluence and comfort. The people of Israel went through the superficial motions of their heritage values but in reality had become the humanists of their age. Great blood-thirsty imperialistic nations lurked on the borders, but the people of Israel would not be distracted from their personally pleasing life style.[6]

In using the prisms of history, philosophy, and theology, Lawrence Patton McDonald was convinced by evidence and experience that the Soviet Union was engaged in a total war against Western civilization that no amount of wishful thinking or appeasement could contain. "It's an economic war," he wrote, continuing,

> It's a war of subversion. It's a war of espionage. It's a war of ideas. And it's a war of terrorism and infiltration. The type of war we are in is far more sophisticated than an exchange of gunfire or nuclear weapons, even, because it's a war of attack upon institutions; it's a war of attack upon every segment of society. It is total war.[7]

The Soviet's destruction of an unarmed civilian airliner with McDonald and 268 other men, women, and children aboard illustrated a graphic and brutal aspect of this war. All are now casualties of this war. How it happened, why, and the events that followed are the purpose of this work. In looking at the available evidence, the issues and ideas relating to this tragedy caused by an act of terrorism, it becomes Dr. McDonald's final testament that may help us understand him better in death than he was understood by many during his life.

1

Technological Terror in the Twilight

"The worst enemy of human hope is not brute facts, but men of brains who will not face them."[1]

Max Eastman
1955

In the early morning twilight hours of September 1, 1983, clear skies and a quarter-moon made terrorism by airborne technology a Soviet military-textbook exercise. At thirty-two thousand feet over Soviet-occupied Sakhalin Island, it took less than fifteen minutes to destroy the lives of 29 crew members and 240 sleeping innocent men, women, and children aboard Korean Airlines Flight 007. It can only be imagined what their brief last moments were like.

For the living, the fate of flight 007 is a historic snapshot, a freeze frame of brute facts. In terms of numbers, greater crimes have been committed by the Soviets and their allies. However, the fate of flight 007 and the immediate reaction by the living may be more important than the fact that 269 human beings perished.

What individuals and nations say and do in a time of extreme crisis can reveal their true character and state of mind. The mainspring of human action, or inaction, in a crisis is very often the sum total of a lifetime of true or false assumptions. The statements and subsequent actions of a leadership in a time of crisis are no different. A single event, as history has repeatedly demonstrated, is like a star shell shot into the air, illuminating a larger landscape, revealing for a brief moment realities otherwise hidden in the darkness of deceit.

No doubt exists that the political leadership in the United

States initially regarded the Soviet Union's act of airborne terrorism a hideous high crime of historic importance. The U.S. House of Representatives in its resolution of condemnation stated: "This cold-blooded barbarous attack on a commercial airliner straying off course is one of the most infamous and reprehensible acts in history."[2] The U.S. Senate's resolution was less sweeping, but no less conscious of the act's historic importance. "This cold-blooded attack," it stated, "on a commercial airliner's [sic] straying off course will rank among one of the most infamous and reprehensible acts of aviation history."[3]

President Ronald Reagan on September 5, 1983, in a nationwide televised address described the act as a "massacre," adding that it was a "crime against humanity that must not be forgotten" and was an "act of barbarism."[4] Later, on September 17, Mr. Reagan maintained, "History will say that this tragedy was a major turning point because this time the world did not go back to business as usual."[5]

Nevertheless, this contradicts what President Reagan had said only forty-eight hours after the Soviets had shot down KAL 007. On September 3, with U.S. disarmament negotiator Ambassador Paul Nitze at his side, the president told reporters:

> I don't believe that [the Soviet incident] should reduce the importance of continuing the talks that we hope will lead to a reduction in the number of nuclear weapons in the world. I think peace is that all-important that we shall continue these talks. That doesn't lessen our feeling, our anger about that terrible tragedy, and the Soviet attitude that they've taken following it. But I think we agree the disarmament talks must continue.[6]

A White House reporter received no reply when he asked: "Mr. President, if you say they are barbarians, how can we negotiate?"[7]

A Reagan supporter and Republican, U.S. Sen. James McClure of Idaho, had little faith in the Soviets on disarmament before 007 and even less after. Speaking on the Senate floor, he said:

> I believe that the Soviet willingness to violate their own domestic law and international law in order to shoot down an innocent

civilian airliner is fully consistent with Soviet willingness to violate the SALT II Treaty with which they have repeatedly pledged to comply.[8]

Senator McClure's anger and distrust were shared by another Reagan supporter and Republican who was halfway around the world in Seoul when the shooting was confirmed, Sen. Orrin Hatch of Utah. Addressing a conference commemorating the thirtieth anniversary of the Mutual Defense Treaty between the United States and South Korea, Hatch said: "I know from conversations with the State Department that many of [the president's] advisors plan to urge business as usual with the Russians."[9]

Senator Hatch also indicated that business as usual lay behind U.S. Secretary of State George Shultz's refusal to cancel a planned September 9 meeting with Soviet Foreign Minister Andrei Gromyko in Madrid, Spain. Gromyko infuriated Shultz at this meeting by showing no remorse at the monstrous criminality committed by Kremlin fighter planes. And Gromyko vowed to shoot down any other unarmed civilian aircraft that accidentally strayed into Soviet airspace.

The Madrid meeting was also attended by the Western allies who vigorously condemned the massacre at a meeting originally convened to ratify compliance of the 1975 Helsinki Human Rights Accords. The Soviets had signed it, which implies they have been abiding by it, "when they are not doing so and when they have arrested most of the Soviet citizens who have been monitoring the Helsinki agreement,"[10] Hatch said during his same speech at Seoul.

Congressman Lawrence P. McDonald, a Georgia Democrat, had been slated to attend the Seoul conference with Hatch and other U.S. lawmakers and was aboard flight 007. He became the first elected U.S. official to die at the hands of Soviet armed forces. Senator Hatch, with cold rage at the murder of his friend and the 268 others, also revealed that the State Department was urging the president only forty-eight hours after 007 was shot out of the skies to reestablish an agreement on cultural and educational exchanges with Moscow and proceed with talks on a new consulate in New York in exchange for an American diplomatic office in Kiev!

Senator Hatch said with a trace of frustration in his voice:

> My hope is that President Reagan will remember the long, slow slide into World War Two, in which ever more outrageous acts by aggressive tyrants were met with moral condemnation and clucking of tongues while business and diplomacy as usual continued to mark the policies of the free world. Each time we let the Soviets escape punishment for their barbaric acts, we ourselves condone and become accomplices in their behavior.[11]

It passed unnoticed that the Soviet act of airborne terrorism took place on the forty-fifth anniversary of Adolf Hitler's planned seizure of the Sudetenland, now western Czechoslovakia, that led to the infamous Munich Agreement signed by Hitler and British Prime Minister Neville Chamberlain on September 30, 1938. Munich not only led to the Nazi seizure of all Czechoslovakia six months later, but to the 1939 Nazi military alliance with the Soviets and the German-Soviet invasion of Poland and the mass murders of World War II.

British historian A. L. Rowse, in his study of British and Western appeasement of Hitler, maintained that the political, intellectual, and moral leaders of the West

> really did not know what they were dealing with, or the nature and degree of the evil thing they were up against. To be so uninstructed . . . was in itself a dereliction of duty.
>
> They would not listen to warnings because they did not wish to hear. And they did not think things out, because there was a fatal confusion in their minds between the interests of their social order and the interests of their country. They did not say much about it, since that would have given the game away.[12]

Congressman Lawrence McDonald's nine-year career in the U.S. Congress was a case study of issuing warnings about the mortal danger of Soviet global designs and having them ignored. As a medical doctor raised in a family of physicians, his entire life and training were governed by examining the evidence, drawing conclusions from the facts, and prescribing a cure; this method he applied both to his medical and his political life. Because he was as direct and honest as his handsome face appeared and was a leader in the John Birch Society, his persistent warnings over nine years fell on the disbelieving ears of those afflicted with a strange bigotry toward evidence and experience.

Dr. McDonald's friend and fellow congressman from Georgia, Republican Newt Gingrich, reminded the House of Representatives of McDonald's powers of diagnostic political prophecy when it came to the Soviets.

> Larry McDonald understood one great central truth. He knew that the Soviet dictatorship is an evil empire and that Soviet leaders would kill innocent people. Larry McDonald understood that innocent men, women, and children could be killed flying near the Soviet Union. Indeed, if one went back and read Larry McDonald's predictions about the Soviet Union in Vietnam, Cambodia, Laos, Angola, Nicaragua, and compared his predictions to those of the State Department or nationally known political spokesmen in foreign affairs, it was Larry McDonald who tended to be more correct.[13]

One of the last public statements released by the physician-turned-politician concerned Western transfer of technology to the Soviets and other totalitarian Communist nations. He warned that the United States cannot maintain a defense superior to the Soviets if it continues the practice.[14] McDonald had been openly critical of President Reagan's lifting the embargo on pipe-laying technology to the Soviets for the construction of a mammoth Trans-Siberian natural gas pipeline to be built by Russian political prisoners.

The Georgia Democrat saw the Reagan decision as a repeat of the folly and foolishness of the 1930s. Less than three months before his death McDonald wrote:

> Many government officials are enthusiastically promoting trade with the greatest tyranny [sic] the world has ever known: the Soviet and Chinese dictatorships. One imagines that they would have been leading the parade to trade with Hitler in the mid- to late-1930s, apparently unwilling to accept the reality of tyranny and the Nazi's desire for world domination. Today the names have changed. No longer is it Hitler, but rather Andropov and Teng. But the goal is the same: destruction of the West.[15]

In the immediate aftermath of the Soviet midair massacre Congressman Austin Murphy, Democrat of Pennsylvania, said he was angry for reasons that went beyond the barbaric act. He told the House:

> I am angry that American technology sold to the Soviets was the basis for the guidance system, the radar, and the computers used by the Soviet plane. I am angry that our response has been confined to shutting down two Aeroflot offices, temporary suspension of landing rights at U.S. airports, and volumes of rhetoric. Other than that it seems to be business as usual with the Soviets.[16]

The murder of McDonald and 268 others demonstrates in the most direct and personally dramatic way that decisions in Washington can have lethal consequences in Moscow. However, such direct evidence and experience appeared to have little effect on the political leadership in Washington or in some big business boardrooms around the nation. A moral amnesia set in only three weeks after the 007 massacre. One hundred top American agricultural and high-technology company representatives, apparently taking their cue from Washington, announced on September 23, 1983, that they would go ahead with plans to attend an agricultural trade exhibit sponsored by the USSR Trade and Economic Council and the Chamber of Commerce and Industry in Moscow. Ralston Purina, Occidental Petroleum, Monsanto Chemical, Archer Daniels Midland, and Coca-Cola made it known they would go to Moscow for the trade fair despite the midair massacre.[17]

McDonald would not have been surprised or angry, but only saddened that men and women of brains continue to refuse to face the brute facts on which their very lives and survival depend. As a trained physician he dispassionately looked on such behavior as a doctor would look on a patient suffering from a potentially fatal disease who is unwilling to accept either the diagnosis or the prescribed cure as a means for survival.

In July 1977 McDonald told the author:

> As a physician, if you have a sick patient I think you need to examine the whole body and look at all the lab tests, look at all the X rays. I think if you are only going to limit yourself to the

hand sticking from beneath the covers, but leave the rest of the body untouched, you will have a very bad success rate. And I am afraid this makes plain some of the problems we have faced over the past thirty years.[18]

2

Soviet Character in the Skies

***"In Soviet psychology, being the strongest guy on the block is never having to say you are sorry."*[1]**

Director Seymour Weiss
Bureau of Political-Military Affairs
U.S. State Department, 1973–74

Since the end of World War II and the start of the Cold War the record of Soviet behavior in the skies has followed a consistent, almost predictable, pattern.

Between April 1950 and September 1983 the Soviets shot at or downed twenty-nine civilian and military aircraft with the loss of hundreds of lives.[2] Between October 15, 1945, and July 1, 1960—a period prior to the development of sophisticated satellites and electronic intelligence technology—eighty-one U.S. military personnel lost their lives in twenty-seven separate incidents due to the Soviets' downing of military reconnaissance aircraft.[3] On July 1, 1960, for example, a U.S. Air Force RB-47 reconnaissance plane with a crew of eight was shot down over the Kola Peninsula near the Polar Barents Sea by Soviet fighters. Four crewmen perished, and the other four were taken captive. The Kremlin insisted that the aircraft had violated Soviet airspace on a spy mission.

However, Soviet army intelligence officer, Col. Oleg Penkovskiy, revealed in his diaries, which were smuggled out of Russia to the West before he was caught as a spy and shot, that the RB-47 was shot down over international waters on direct orders from Premier Nikita Khrushchev. Almost five years after the incident Penkovskiy wrote:

> Such is our way of observing international law. Yet Khrushchev was afraid to admit what had actually happened. Lies and de-

> ceit are all around us. There is no truth anywhere. I know for a fact that our military leaders had a note prepared with apologies for the incident, but Khrushchev said: "No, let them know that we are strong."[4]

In May 1960, U-2 pilot Francis Gary Powers's plane was shot down over the USSR. Concerning this incident, Soviet scholar Dr. Frederick Barghoorn wrote:

> Khrushchev apparently felt that he had a golden opportunity to blacken the image of the United States as a "warmongering" nation. Soviet propaganda regarding the U-2 affair was utilized to build up the Soviet image of a predatory United States and, at the same time, to terrorize small and weak nations.[5]

In the wake of the KAL 007 tragedy, two decades later, the Kremlin leadership launched a strikingly similar global peace offensive by attempting to prove to the world that KAL 007 was, indeed, a spy plane. The Soviet peace offensive even included staged peace demonstrations in Moscow, aimed at influencing the nuclear freeze movement in the West while clouding the true facts of the Soviets' thirty-five-year history of shooting down military and civilian aircraft. The Soviet peace offensive also shifted public focus away from the Soviet record of regular airborne intrusions over North America and how the United States deals with such deliberate overflights.

From January to September 1983, for example, seventy-seven Soviet or Eastern European Communist air carriers illegally entered U.S. airspace while on nonstop flights from the USSR to Cuba. The Soviet air defense system, smaller than the U.S. North American Air Defense Command (NORAD), has over the years fired as many as nine hundred Surface to Air Missiles (SAM) at Western nations' reconnaissance aircraft without hitting any of their targets.[6] The Kremlin places great urgency on air defense, maintaining a vast network of radar installations, with 2,250 fighter interceptors and as many as ten thousand SAM missiles as anti-aircraft defense.[7]

Until the United States temporarily suspended landing rights in 1981 as punishment for the Soviet Union's aiding and abetting repression in Poland, Soviet commercial Aeroflot flights deliberately departed from regular U.S. inflight routes to photograph sensitive defense installations. Two Aeroflot planes, a Polish, and a Czech air carrier are known to have passed over the naval shipyard at Groton, Connecticut, where work was underway on a nuclear submarine.[8]

Deputy U.S. ambassador to the U.N. Security Council, Charles M. Lichenstein, during the Security Council debate over the KAL 007 massacre, revealed that he had reviewed the log of seventy-five instances of Aeroflot air intrusions into American airspace. On November 8, 1981, he reported, an Aeroflot plane departed from its air route over open water to fly over the Navy base at Groton on its way to Dulles International Airport near Washington.

Ambassador Lichenstein told the Security Council:

> Several days later, the same aircraft, on leaving Dulles for its return flight, flew a similar unauthorized route over New England. My government lodged a very firm protest. My government then imposed what it considered a proportionate penalty. It suspended Aeroflot scheduled service into Dulles for two flights. It did not authorize the use of a heat-seeking missile![9]

Bruce Herbert, deputy director of the Washington-based Center for International Security, a military and intelligence strategy "think tank," maintains that the Soviets violate U.S. airspace sometimes as much as a hundred times a year. Herbert, who is also a U.S. Naval Reserve transport pilot, also maintains that Bear and Bison Soviet aircraft based in Cuba overfly the United States for the purpose of what is termed "fence checking," testing of our radar to determine how long F–14 fighters take to intercept an air intruder.[10]

Herbert points out that the Soviets, while insisting on the sanctity of their own borders and territorial waters, are "routinely hypocritical" about intruding Western airspace and sending submarines into territorial waters of other nations. Soviet diplomats caught spying are regularly expelled from free and not-so-free nations.

Kremlin Foreign Minister Andrei Gromyko underscored this double standard in his statement that defended the downing of KAL 007. "Soviet territory, the borders of the Soviet Union, are sacred," he said without equivocation. "No matter who resorts to provocations of that kind, he should know that he will bear the full brunt of responsibility for it."[11]

Senator Frank Murkowski (R–Alaska) revealed in the wake of KAL 007 that since 1978 the number of Soviet aircraft invading the airspace over Alaska is about 50 percent higher than the number of similar intrusions over the entire continental United States during the same period.[12] "United States policy," he stated, "is to safely escort even military intruders out of American airspace, while the Soviets shoot down aircraft lost and in distress."[13]

A half-million Alaskans live near the Soviets; some are only two nautical miles from the USSR. During 1983, up to the time that KAL 007 was destroyed, fourteen Soviet aircraft had invaded the skies over Senator Murkowski's home state, and they do so about fifteen times a year. Murkowski added:

> In the last five years Soviet aircraft have been escorted out of U.S. airspace in Alaska by American interceptors roughly eighty times. In fact, in the years 1963, 1968, 1969, 1974, and 1983, Soviet military aircraft not only penetrated U.S. airspace, but actually flew over Alaskan soil. Few Americans know that in February 1974, an amphibious Soviet coke aircraft actually landed at Gamble on St. Lawrence Island off the west coast of Alaska.[14]

While the Soviets routinely penetrate North American airspace with impunity, the United States' ability to protect the skies over the Aleutian Islands chain off the coast of Alaska has not improved since World War II when the Japanese invaded and captured Attu and Kiska islands. At least that was the contention of Lt. Gen. Bruce K. Brown, commander of the Alaska Air Command, eight days after the shattering news that KAL 007 had been shot down. "If they wanted to," Brown said of the Kremlin leaders, "they could go all the way and take Omaha and never be detected."[15]

Omaha is headquarters for the Strategic Air Command (SAC). Brown was former vice-commander of the NORAD in Colorado and is in a position to judge the strengths and weaknesses of American air defenses. He insisted in the wake of KAL 007 that NORAD early warning stations in Alaska are linked by a single Alascom satellite communications system and, should it go dead or be knocked out, the United States would be blind to a Soviet surprise first strike. "The whole country's that way," he bluntly stated. "It's been that way for fifteen years. We've been telling you that, and nobody wants to hear it. Nobody wants to hear about more money on defense."[16]

Seymour Weiss, director of the State Department's bureau of political and military affairs in the early 1970s, pointed out that the attitude of the Kremlin leaders and that of Soviet Foreign Minister Gromyko over the 1983 Korean Airline tragedy is consistent with their past behavior. When Soviet submarines penetrated Scandinavian waters they were unapologetic, even hostile and belligerent; the Kremlin's invasion of Afghanistan was accompanied by no effort to save face; and massive infusions of military arms to Syria and Central America by the Soviets are neither denied nor hidden. The shooting down of KAL 007 "may reveal an emerging overt and considered policy to flex its military muscle." Weiss added:

> It is not implausible to suspect that the Soviet leadership is becoming increasingly aware that its massive effort of the past two decades to surpass the U.S. military in both nuclear and conventional strength ought properly to have a political payoff. In the Soviet psychology, being the strongest guy on the block is never having to say you are sorry.[17]

During his nine years in Congress and as a member of the House Armed Services Committee, Congressman McDonald had repeatedly warned that neglecting national defense and accepting the Kremlin's professions of peaceful intent were dangerous snares and delusions.

McDonald's House colleague, Congressman Carroll Hubbard, Democrat from Kentucky, was slated to board KAL 007 for the trip to South Korea, but a local speaking date in his district caused him to change his plans and take a different plane. In the aftermath of KAL 007, Hubbard told the House:

> By his death Larry McDonald causes many more people to be aware of the brutal, cruel attitude of the Soviet Union's lead-

ership and military than he ever did by his words on the floor of the U.S. House of Representatives. In other words, Larry McDonald tried to convince us during his lifetime that the Soviet Union regarded their airspace to be of higher priority than human lives, even those inside the Soviet Union. Now that Larry McDonald is dead, we know he was right.[18]

3

Dress Rehearsal for Disaster

***"I wonder if the free world does not bear some responsibility for the deaths of 269 civilians by our collective failure to react to the Soviets' 1978 attack."*[1]**

U.S. Rep. Jack Kemp (R–N.Y.)
September 14, 1983

"If the United States is really serious about human rights," wrote the editors of the *Wall Street Journal* on April 25, 1978, "it should make a louder noise about the fate of the 108 travelers of flight 902."[2]

In the early afternoon of April 20, 1978, air traffic control at Orly Airport outside Paris cleared Korean Airlines Flight 902 for take-off. The Boeing 707 was quickly airborne, bound for Seoul via the Polar route.[3] Thirty-six-year-old Seoul salesman Bahng Tai Hwang glimpsed outside from his center seat while Paris disappeared from view. Thirty-one-year-old Yoshitako Sugano was looking forward to returning to his prosperous coffee shop business in Yokohama, Japan.

Neither Bahng nor Yoshitako would ever see home again. KAL 902 strayed into Soviet airspace, and eighteen minutes later a Soviet fighter interceptor fired three cannon bursts into the plane. Bahng died instantly, and Yoshitako bled to death. Thirteen other passengers suffered major to minor wounds.[4]

The pilot of KAL 902, Captain Kim Chang Kyu testified that, besides the cannon bursts by the Soviet interceptor, his plane was hit by a missile that sheared off fifteen feet of the 707's left wing. "There was smoke everywhere," he recalled. "The emergency bells were ringing and the plane was bucking and rocking all over. I felt like I was on a runaway horse."[5]

Captain Kim put the stricken airliner into a steep dive, for he knew that if he did not reach a lower altitude his passengers might all die from a loss of oxygen within a few minutes after losing cabin pressure at thirty-two thousand feet. Also, at that altitude the temperatures were well below minus 40 degrees Fahrenheit. With extraordinary skill he managed to regain control of the 707 after descending to a safer altitude of three thousand feet. After burning up fuel for over an hour to minimize the risk of fire during landing, he cooly brought down KAL 902 on a frozen lake near the Russian town of Krem, south of Murmansk and about four hundred miles northeast of Leningrad.[6]

In September 1983, Captain Kim observed that in almost all details the Russians made the same claims about KAL 902 that they have about KAL 007. "After I was shot down," he noted, "the Russians made the same claims we're hearing now. They said, 'We tracked you for more than two hours, flew around the plane, fired tracers in front of you'—all that. It all sounds exactly the same this time."[7]

In the U.N. Security Council debate over KAL 007, U.S. Ambassador Jeane Kirkpatrick noted the similarities between the two air tragedies, including the Soviets' telling one version and, in the case of KAL 902, Captain Kim's telling another. Kirkpatrick told the Security Council:

> He tells us he saw the plane only once, off to the right and somewhat behind him. He thought this was strange, since international guidelines call for intercepting fighters to fly to the left of the plane, where the pilot sits. When Mr. Kim's copilot, who had a clearer view of the plane, reported that it bore the red Soviet star, Mr. Kim immediately slowed his speed and turned his landing lights off and on repeatedly, the recognized international signal that an aircraft will follow the interceptor's directions. In addition, Mr. Kim tried to establish contact with the Soviet craft, but the two planes' radios were on different frequencies. In any event, the next thing Mr. Kim knew, a missile fired by the Soviet pilot had torn off a good part of his plane's left wing.[8]

As with KAL 007, the Soviets' attack on KAL 902 was in darkness, with the pale polar Arctic skies providing clear visibility for Soviet fighters to see that it was an unarmed commercial airliner. Initially, it was reported by the Soviets that the aircraft had been forced down. Not until the 108 passengers and all but two KAL crew members were flown to the safety of Helsinki, Finland, did the world discover that the Soviets had shot at and damaged the plane, which forced it to land.[9] The Soviets put out the story through Western journalists stationed in Moscow that orders *directly* from the Kremlin were given for the missiles to be fired on the KAL 902 because it was suspected of being an intelligence spy plane.[10]

From the time of the KAL 902 incident to the aftermath of KAL 007, it was accepted by the West that in both cases Soviet fighters stalked each aircraft for two hours or more. In the case of KAL 902, however, the Soviet claim is contradicted by a detailed diary of the ordeal kept by a Japanese passenger and by the testimony of Captain Kim and his crew.

According to the diary of forty-two-year-old Seiko Shiozaki, only one hour and forty-two minutes elapsed between the time the Soviet fighter crippled the aircraft and when it landed on the frozen lake.[11] It was only eighteen minutes after the Boeing 707 had entered Soviet airspace when the plane was attacked, according to radio intercepts. Yet the world had been led by the Soviets to believe that the KAL aircraft wandered in Soviet airspace for some two hours prior to their "interception" by a missile.

Navigation error, common in the polar region, was offered to explain KAL 902's overflying Soviet airspace and sensitive naval installations in Murmansk. However, after their release several passengers said they had become concerned when they noticed that the plane, which had been flying toward the setting sun, made a near 180-degree turn.[12] After the KAL crew safely returned from Russia, all testified that just prior to becoming lost "an electrical shock paralyzed the navigation system."[13]

In the five and a half years between the KAL 902 incident over Murmansk and the KAL 007 tragedy over Sakhalin Island at the other end of the Soviet Union, no explanations or even theories have been advanced publicly for the "electric" shock to KAL 902's navigation system, nor could the instruments be checked since the USSR refused to return the plane or instruments to the West.

However, in the aftermath of KAL 007, veteran foreign correspondent Hilaire du Berrier revealed the substance of a conversation he had had in Copenhagen on April 20, 1978, with the South Korean ambassador to Denmark, Chang Chi Ryang. The former air marshal in the South Korean air force is reported to have communicated with his Soviet counterpart in Copenhagen on the day that KAL 902 was shot down. Du Berrier quotes the South Korean envoy as saying to the Russian:

> I am myself an aviation man. I know that the Soviet Union has perfected a means of throwing off an airplane's navigational equipment and that it picked a South Korean plane for testing its new device, knowing that South Korea is too tiny to retaliate against the mighty Kremlin. Nevertheless, you are going to release our crew and passengers, and I will say nothing of the matter nor will you.[14]

Two days later the passengers and crew were set free and flown by American commercial airlines to Helsinki.

Chang insists that KAL 007 was drawn off course in the Far East by the same kind of electronic device as KAL 902 in an act of terrorist intimidation. The similarities in the two cases are remarkable and cannot be simply chalked off as two separate "incidents" or errors, or as a coincidence.

Both commercial airliners experienced unusual and unprecedented navigational troubles; both experienced savage Soviet response; both were charged after the attacks with being spy planes; and in both cases the Soviets spread a blanket of falsehoods to justify their true crimes. In KAL 902's case the Soviets seized the aircraft flight recorder and cockpit recorder and would not release them; and in KAL 007's case, the Kremlin spent extraordinary time and manpower trying to retrieve them from the wreckage in the waters off Sakhalin Island.[15]

The parallels in the two incidents even extend to the field of diplomacy and how the United States reacted in the aftermath of each. President Carter's secretary of state, Cyrus Vance, was in Moscow when KAL 902 was shot down, discussing arms reductions with Soviet Foreign Minister Andrei Gromyko.[16] President Reagan's secretary of state, George Shultz, was slated to meet Gromyko in Madrid, Spain, when KAL 007 was shot down by the Soviets. In each case the Soviets were belligerent, while, paradoxically, launching a peace offensive and seeking to terrorize the world with the threat of nuclear conflict.

During the intense and emotional denunciation of the Soviets' downing of KAL 007, members of the Senate and House ignored the 1978 dress rehearsal for disaster. The lone exception was Rep. Jack Kemp (R–N.Y.). Lamenting the loss of Lawrence McDonald and the 268 others who died with him, Kemp went on to warn that a resolution of condemnation was not enough:

> We would dishonor, rather than honor, the memory of those who died aboard the Korean airliner if we allowed our response to stop there.
>
> I wonder if the free world does not bear some of the responsibility for the deaths of 269 civilians by our collective failure to react to the Soviets' 1978 attack. I am convinced that a resolution of disapproval, standing alone, will not influence future Soviet behavior one whit.[17]

Within two days after Kemp's words a watered-down resolution was all the nation received from their elected representatives in Washington.

McDonald would not have been surprised. He had said that one of the many problems with most members of Congress is that they do not have a historical perspective.

"The great use of history," he wrote in January 1981, "is to recognize in it recurring patterns which, read rightly, may serve as guideposts for action. History does not have to repeat itself if persons who understand it take proper precautions."[18]

4

Barbarians with Ballistic Missiles?

". . . the present Soviet leaders began their careers at a time when the regime was practicing massacre on a grand scale."[1]

Robert Conquest
British Historian-Poet
September 11, 1983

During the U.S. Senate debate over the Soviet destruction of KAL 007, a historical parallel was drawn by Alaskan Republican Frank Murkowski between the brutal actions of the Soviet leaders in the twentieth century and the czarist rulers of the eighteenth century. He told the Senate:

> The forefathers of Alaskan natives personally experienced the brutality of Russian intrusions into their land and lives as early as the eighteenth century. Many Alaskan natives gave their lives while resisting Russian naval attacks, colonialism, and the cruelty of the Russian trapper who settled in Alaska.[2]

Does the historical evidence support the view that a continuity of cruelty and barbaric behavior exists throughout Russian history, beginning with the invasion by the Mongol hordes in the thirteenth century and the three-hundred-year Tartar occupation and continuing throughout the reign of the czars to the present dictatorship? Moreover, does a parallel exist between czarist global ambitions and those of the modern Soviet state?

Alexis de Tocqueville, a French aristocrat who journeyed through the United States for nine months between 1832 and 1833, spelled out the differences between czarist Russia and the United States in his classic work, *Democracy in America.* With remarkable accuracy de Tocqueville correctly forecast that the two powerful nations of the future would be the United States and Russia.

> The American struggles against the natural obstacles which oppose him; the adversaries of the Russian are men; the former combats the wilderness and savage life; the latter, civilization with all its weapons and its arts; the conquests of the one are therefore gained by the ploughshare; those of the other, by the sword. The Anglo-American relies on personal interest to accomplish his ends and gives free scope to the unguided exertions and common sense of the citizens; the Russian centres all the authority of society in a single arm: the principal instrument of the former is freedom; of the latter, servitude. Their starting point is different, and their courses are not the same; yet each of them seems to be marked out by the will of Heaven to sway the destinies of half the globe.[3]

In the same decade that de Tocqueville penned his prophetic work, another French aristocrat undertook a journey through Russia in 1839. The Marquis de Custine's journal is not as well known as *Democracy in America,* but it is every bit as prophetic.

The modern English-speaking world did not become aware of Custine's *Journey For Our Time* until 1951, when the wife of U.S. diplomat Foy D. Kohler translated a copy. She had found it by accident in a second-hand Moscow bookstore when her husband was posted to Moscow in the years shortly after the end of World War II. Phyllis Penn Kohler maintains that through Custine's journal we gain an insightful look into the Russia of both the czars and the commissars, a continuity that is as uncanny as it is cruel.

In 1930 the Soviet regime "discovered" Custine's book and decided to publish a Russian edition as part of a continuing campaign to discredit its czarist predecessors. Mrs. Kohler wrote:

> It seems that the new rulers in the Kremlin soon realized that Russian readers were viewing Custine's observations as decidedly pertinent to Bolshevik, as well as czarist, despotism, and ordered all copies confiscated.[4]

Fully one hundred and thirty years before Nobel Prize-winning Russian novelist Aleksandr Solzhenitsyn delivered his searing indictment of the barbarism of the Soviet system as one great Gulag, Custine wrote:

> It can be said of the Russians, great and small—they are intoxicated with slavery. . . . They wish to rule the world by conquest; they mean to seize by armed force the countries accessible to them, and thence to oppress the rest of the world by terror. The extension of power they dream of is in no way either intelligent or moral; and if God grants it to them, it will be for the woe of the world.[5]

Custine's biting commentary of Russia was not aimed at Czar Nicholas I, whom he personally liked and even admired. Rather, the French aristocrat, who had gone to Russia a convinced monarchist, took issue with the *system.* What appalled him was its brutalizing influence, the Russian people's submission to slavery and the czarist system. The czar was able to maintain that system by stimulating the appetite of his enslaved subjects for acquiring glory, power, and wealth—looking on other lands as fit objects for enslavement and plunder.

"Either the civilized world will, before fifty years have passed," Custine predicted, "fall again under the yoke of the barbarians, or Russia will undergo a revolution more terrible than the revolution whose effects are still felt in Western Europe."[6] He was referring to the French Revolution, which Lenin would use as a model.

Custine was only off in his prediction by fifteen years, since the Russian upheaval of 1905 proved to be a dress rehearsal for the Revolution of 1917, launched after the czarist armies had been demoralized and decimated by World War I.

Custine, a classicist like de Tocqueville, understood more profoundly than anyone else of his generation the tragedy that had befallen Russia. The lies, deceits, secrecy, and brutality he saw firsthand were, he believed, the products of the Mongol invasion (1237–1240) and occupation of Russia for three centuries, which crippled its capacity to evolve from barbarism to civilization.

Custine wrote:

> Barely free of her foreign yoke, she [Russia] thought that everything that was not Mongolian domination was freedom; it was

> thus that in the joy of her inexperience she accepted servitude as a deliverance, because it was imposed on her by her legitimate sovereigns. This people, debased under conquest, felt happy enough, independent enough, provided its tyrant was called by a Russian instead of a Tartar name. . . .[7]

Lenin's seizure of power and his establishment of the Soviet dictatorship, which his heir Josef Stalin would refine at the cost of millions of lives, were repetitions of Russia's historical tragedies and errors. The crimes of Lenin, Stalin, and their successors were evaded or rationalized by many in the West because, as during the reigns of Peter and Catherine, evil acts were done in the name of doing good. Russia had exchanged the tyranny of the czars for the tyranny of the commissars.

Former U.S. ambassador to the Soviet Union Foy D. Kohler pointed out that Stalin admired Ivan the Terrible (1530–1584). The career diplomat saw little difference between the suppression of freedom in Poland in 1832 by Nicholas I and the crushing of Czechoslovakia in 1968 by Brezhnev and Kosygin. Kohler noted:

> Today, as in Custine's time, Russia is ruled by a small minority. . . . The czars, however, ruled with the legitimacy of the accepted doctrine of divine right of kings. Today's rulers simply seized power by force and have since preserved that power by an all-pervasive dictatorial system.[8]

A few months before the 1968 Soviet invasion of Czechoslovakia, the British scholar, historian, and poet Robert Conquest published his definitive work, *The Great Terror,* a study of Stalin's political purges in the Soviet Union during the 1930s. "It was launched," wrote Conquest, "in the coldest of cold blood, when Russia had at least reached a comparatively calm and even moderately prosperous condition."[9]

In the immediate aftermath of the KAL 007 massacre, Conquest expressed surprise that so many in the West were shocked by the brutality and deceit of the Soviets. He concluded that many in the West project their own values on the Soviets and assume they must be shared by the Kremlin leaders. He also noted that the airline tragedy happened on the fiftieth anniversary of collectivization by Stalin of Soviet farms in the Ukraine, which created the only known man-made famine in history (14 million perished).

Conquest wrote:

> It is alien to our political culture to consider that there are political rulers who really do not mind killing people. A glance at the history of Tamerlane and Genghis Khan should be enough to remove that delusion. And the present Soviet leaders began their careers at a time when the regime was practicing massacre on a grand scale.[10]

The American scholar Dr. Richard P. Clayberg, who has made the study of Soviet psychology his specialty for fifteen years, echoes Conquest. In the face of the enormous evidence of past Soviet crimes against the Russian people, he argues, the West has no right to be surprised that they would shoot out of the skies in the dark of night an unarmed civilian airliner with 269 people aboard.

Dr. Clayberg points out, moreover, that the Soviets are compulsively paranoid about spies and their territory because, as a tyranny, they have much to hide. They maintain a compulsive secrecy that dates back many centuries.

> It is a system that is structured against thinking and reinforces through its educational system obedience to the memorization of dogma. It is a system that breeds its own ignorance, it breeds its own barbarians, it breeds its own insensitivity, and it breeds its own paranoia.
>
> The Soviet system is rigid, not stable. Stability implies acceptance of the values of a system as it is. But rigid means you go along with the system, one that depends on and is sustained by informers, lies, and deceit. What you get is a gangster regime that hates us not for what we do, but for what we are. President Jimmy Carter had a terrible time learning that bitter lesson. There are other people in the West who will never learn.[11]

Custine in his own time worried about Russians with weapons in their hands. We in our time worry about Soviet leaders armed with ballistic missiles. The efforts at détente by the West in the 1970s and President Carter's call in 1977 that we should get over our "inordinate fear" of communism did little to stay the hand of the Soviets from invading Afghanistan and from crushing the Polish Solidarity movement, and did nothing to stay their brute resolution to shoot down KAL 007 with 269 civilians aboard. Such behavior illustrates that trade, expanded cultural contacts, and efforts at détente do little to influence Soviet actions. In fact, a case can be made that the Kremlin leaders, like the Oriental despots of old, look on such policies as both a sign of weakness and a form of tribute.

In July 1977 Congressman McDonald looked skeptically on President Carter's efforts at détente with the Soviets. In a radio network interview he told this author:

> *Détente* is a word with many meanings. It is the kind of Doublethink language the Soviets love because it means one thing to us and another thing to them. To us it means a relaxation of tensions, trade, and increased agreements. To the Soviets it means a mechanism on the road to world conquest.
>
> In the French definition of the word *détente* it has a second meaning: the word *trigger*. No thinking person who has studied this area is willing to state that the Soviets have ever rejected their intention of world domination and world conquest and the elimination of all opposition to communism. . . . Hence, we get into a great deal of difficulty when we delude ourselves into the belief we are both working for détente.[12]

John Barron, author of a definitive study of the KGB (the Soviet secret police), maintains that while the Kremlin leaders insist on "détente" with the West, their own KGB is working to undermine it.

> While the Soviet leadership tries to negotiate a relaxation of tensions with the West, the KGB plans to sabotage Western cities; encourages civil strife in America; plots to incite civil war in Mexico and Ireland; nurtures the Palestinian guerrillas in their worldwide terrorism; strives to corrupt and subvert Western officials and politicians; and, through a variety of deceits, vilifies the same nations whose economic and political favors the leadership courts.[13]

The chairman of the KGB for fifteen years prior to becoming Soviet premier in November 1982 was Yuri Andropov. "The more one learns about Yuri Andropov," McDonald told the House of Representatives five months before his death, "the more uneasy one becomes. As more details of his early career become available, his personality emerges as completely ruthless."[14]

5

The Cobra in the Kremlin

***"If he had tried to have the Pope killed, he would not hesitate to do the same thing to any of them who got in his way."*[1]**

Yuri Andropov: A Secret Passage into the Kremlin
1983

On May 13, 1981, Turkish-born terrorist Mehmet Ali Agca shot Pope John Paul II in Rome's St. Peter's Square. Forty-eight hours later this author spoke with Kasim Gulek, the editor of *Ankara,* Turkey's largest daily newspaper. Gulek made a series of startling statements:

> The attempt on the life of His Holiness is not some plot of rightists as your media has been reporting. Turkish intelligence has grounds to believe this assassin was trained in Bulgaria or some other Eastern European country with the knowledge and approval of the KGB in Moscow. We have every reason to believe the motive for the attempted assassination is related to the Polish situation and the fact that Pope John Paul is a Pole.[2]

Gulek is the former vice-premier of Turkey and is on an intimate first-name basis with members of the Turkish military who, before the shooting of the Pope, had seized power in Turkey to stem a growing nationwide wave of terrorism in that country. The news media in Washington, though made aware of Gulek's assertions, chose to ignore the revelations.

In *The Andropov File,* U.S. historian of Soviet and Chinese affairs Martin Ebon observed that

> Agca, who had escaped from a Turkish prison, spent considerable time in Bulgaria before traveling elsewhere in Europe and going to Rome. Speculation that the Bulgarian secret police, at the behest of the KGB, had financed and instructed Mehmet Ali Agca, was heightened by the fact that, at the time, Yuri Andropov was chairman of the KGB.[3]

Sixteen months after the attempted assassination of Pope John Paul II, Claire Sterling in her landmark work *The Terrorist Network* revealed that Agca was not some neo-Nazi right-wing crackpot working alone, as originally reported by the international news media. In September 1982 she wrote:

> As I learned in months of investigation, there is hard evidence that Agca was an instrument in an elaborate international plot. Whether through negligence, nearsightedness, or indifference, not a single country concerned has pressed an investigation to the end.
>
> No secret police organization has more intimate links with the KGB than Bulgaria's. What is more, the KGB keeps tabs on all terrorists as a matter of course. It is inconceivable that the KGB would not have known all there was to know about a terrorist as closely involved with the Bulgarian secret service as Agca was.[4]

In the same month that Sterling's revelations appeared in print, a TV documentary, "The Man Who Shot the Pope—A Study in Terrorism," was aired. Marvin Kalb and Bill McLaughlin of NBC News traced, after a nine-month investigation, what they called an unbroken trail from Agca to organized crime elements in Turkey to the Bulgarian secret service and the Soviet KGB. "A Soviet connection is strongly suggested, but it cannot be proved," stated Kalb. "It seems safe to conclude [that Agca had] been drawn into the clandestine network of the Bulgarian secret police and, by extension, the Soviet KGB."[5]

Kalb and McLaughlin, and Sterling, in their separate investigations, also maintained that the motivation for attempting to murder the Pope pointed directly to John Paul II's deep involvement in the Polish Solidarity movement. Kalb stated:

> NBC News has learned that in early August [1980], as the crisis escalated, the Pope sent an envoy to the Kremlin whom we are pledged not to identify. He delivered an extraordinary handwritten letter, in Russian, from the Pope to the Soviet leader Brezhnev. It said that though the Pope was the head of the universal church, he was still a Pole, and deeply affected by the developments in Poland. And if the Russians moved against Poland, he would lay down the crown of St. Peter and return to his homeland to stand shoulder to shoulder with his people.[6]

The NBC-TV program quoted former KGB agent and defector Vladimir Sakharov to the effect that the information known to the Bulgarians would also be known to the KGB in the Kremlin. The papal letter to a then gravely ill Brezhnev, who would die less than two months after the airing of the NBC-TV documentary and be replaced by Andropov, "instigated a secret shuttle mission between Moscow, Rome, and Warsaw that led eventually to a temporary easing of the Soviet attitude toward Solidarity."[7]

The Russian-born authors Vladimir Solovyov and Elena Klepikova, in their 1983 profile of Andropov, conclude that the former KGB chief was already running the Soviet Union before Brezhnev died. And they maintain that Andropov had to know of the plot against John Paul II. " He was the chief planner of the pacification of Poland, in spite of all the monkey wrenches thrown into the machinery by his party colleagues. . . . Hence he did not consider it necessary to increase the number of those privy to the plot against John Paul II."[8]

Alex Alexiev, a Soviet and Eastern European affairs analyst with the U.S. government funded Rand Corporation, a U.S. strategy "think tank," concluded that John Paul II's May 1979 visit to Poland led to the rise of the Polish-Catholic Solidarity labor movement. The movement had a politically unsettling impact not only on Poland but also on other Eastern European countries where Catholics make up a significant percentage of the population. Even the Ukrainian and Lithuanian Catholic Church of the Eastern Rite was influenced by Solidarity's strongly Catholic identification.

The Kremlin was fearful that what was happening in Poland would spread to its entire Soviet empire. Alexiev points out that the Soviets, through the KGB headed by Andropov, launched a particularly vicious propaganda campaign against the Pope and that the assassination "marked another watershed in Moscow's campaign against him," and that the propaganda campaign against the Pope continued after he survived the assassination attempt.

Alexiev wrote:

> For Moscow, John Paul II was and continues to be much more than a narrow Polish problem. The Pope on his part realizes clearly what a serious challenge he presents for the Kremlin. As early as 1979 during his visit to Poland he told an audience: "I'm sure there are people out there who are already having a hard time taking this Slavic pope." Hard enough, it seems, to prompt an exasperated Kremlin to cry out as Henry II once did regarding the archbishop of Canterbury: "Will no one rid me of this turbulent priest?"[9]

The attempt to murder the Pope, while horrifying to the entire world, nevertheless resulted in a reluctance on the part of Western governments to pursue an investigation, since the evidence seemed to point directly at the Kremlin and the Soviet KGB. For example, five months after the attack on the Pope, New York Republican Sen. Al D'Amato began his own inquiry by personally going to Rome after it appeared that neither the newly elected Reagan administration nor the Italian government was prepared to conduct a full investigation.

D'Amato told this author in a television interview in February 1983:

> I said nothing in public and turned over to the CIA what I had found, and waited. I waited for eleven months and nothing happened. Vatican officials I had met relayed to me the strong belief that the Soviet Union, fearing John Paul's influence in Poland, had participated in a plot to kill the Pontiff. I relayed to the CIA this information.[10]

In the wake of Kalb and McLaughlin's disclosures on NBC and Mrs. Sterling's in the *Reader's Digest* of September 1982, D'Amato went public for the first time, saying only that their investigations confirmed what he had found in October 1981. The Italian government then launched its own official inquiry and collected substantial evidence that there was a Bulgarian-KGB connection and arrested several suspects. Later, after the death of Brezhnev and the ascendency of Andropov (in November 1982), D'Amato charged that the CIA had failed to pursue the probe in conjunction with the Italians. "It's shocking to me," he said on returning from Rome on February 10, 1983, "that the CIA has shown itself to be either inept or deliberately obstructing the investigation."[11]

Soviet-Chinese affairs expert Martin Ebon quotes former U.S. National Security Advisor Zbigniew Brzezinski and former Secretary of State Henry Kissinger as saying that the assassination attempt on the Pope was probably inspired by the KGB. Why, therefore, were the Reagan administration and the CIA reluctant to pursue the trail first marked out by Kasim Gulek and later by Senator D'Amato, Marvin Kalb, Bill McLaughlin, and Claire Sterling?

Ebon wrote:

> Eventually, any speculative link of the former KGB chief to shots on St. Peter's Square was likely to be forgotten. As official Washington was hoping to establish a cordial, or at least businesslike relation with Andropov, anything linking him to the plot had to be regarded as, to say the least, awkward.[12]

Shortly after the attempt on the life of Pope John Paul II, Congressman Larry McDonald raised publicly the question of KGB involvement in the assassination attempt. As with so many other issues he raised in the House of Representatives during his congressional career, McDonald was ignored. In the wake of the NBC and Sterling disclosures and D'Amato charges, McDonald, in letters to President Reagan and CIA Director William Casey, demanded full disclosure of what they knew about the possibility of a KGB-Bulgarian connection in the assassination plot. McDonald also asked that Rep. Edward Boland (D–Mass.), chairman of the House Select Committee on Intelligence, undertake public hearings.

McDonald observed:

> One must ask whether a deliberate attempt is being made at the highest levels of our government to suppress the KGB's involvement in the assassination attempt to preserve nonexistent "détente," arms control talks, and trade. . . . For officials of our government to suppress evidence of such a horrible crime would be immoral and put us in the same category as the nation that plotted the assassination. It would be detrimental to our own national security.[13]

In the nine months between Andropov's formal installation as Soviet premier and the destruction of KAL 007 with McDonald aboard, Soviet and Western behavior provided sharp contrasts. For example, the *Wall Street Journal*, in a survey of top U.S. Catholic bishops and lay leaders conducted at the time McDonald raised the issue of a cover-up, revealed a strange attitude toward the attempted assassination. The *Journal* quoted Catholic John Cardinal Krol of Philadelphia as acknowledging the Soviets' drive for global domination and that they may well have been behind the attack on Pope John Paul II. But, Krol said, this should not prevent U.S. efforts to reach arms control agreements with the Kremlin![14]

In an editorial of January 2, 1983, the *Wall Street Journal* pointed out that, in contrast, Soviet news agency Tass, in replying to charges that the Kremlin was linked to the papal plot, inferred that "the Pope got what he deserved" by his "subversive activity" among Catholics in Eastern Europe.

> While Western intelligence struggles to avoid implicating the Communist world in the shooting of the Pope, Moscow nonchalantly admits to the motive. Far from distancing itself from aggressive designs on the pontiff, it associates itself with them.[15]

Recalling that Hitler in the 1930s used the alternating tactics of terrorist threats and peace-seeking overtures, the *Journal* concluded that the Soviets' use of poison gas in Afghanistan, rumors of Soviet violations of strategic arms agreements, "and now reading Tass on the Pope, we have to worry that Mr. Andropov is playing a similar hand today, and wonder how far he may get."[16]

Andropov had come a long way since he was USSR Ambassador to Hungary during the anti-Soviet uprising in 1956 when, in the words of John Barron in his study of the KGB, "he demonstrated a first-rate capacity for intrigue by helping lure Hungarian leaders to their deaths."[17] His fifteen years as chief of the KGB gave him the means to consolidate his position that paved his way to the top Kremlin post of power.

Soviet authors Vladimir Solovyov and Elena Klepikova maintain that Andropov's link with the papal assassination attempt signaled a return to the Stalinist era, with profound implications for those in the West as well as those in the Kremlin.

> None of [the Kremlin leaders] had any concern for the Pope; they were thinking only of themselves and of their total defenselessness before that enigmatic man with the ambiguous smile on his thin lips and the nearsighted eyes behind his thick glasses. If he had tried to have the Pope killed, he would not hesitate to do the same thing to any of them who got in his way. . . . Beginning on May 13, 1981, the Stalinist terror returned to the Kremlin.[18]

In the remaining months of his life, McDonald filled the *Congressional Record* with long and detailed accounts of Andropov's ruthless record during his fifteen years as KGB chief. Dismissing as "bunk" early press accounts of Andropov's requiring time to consolidate power and of his liking for Western music and art, McDonald saw him as a consummately ruthless policeman, a deadly cobra who mesmerized his victims before striking.

McDonald told the House of Representatives on February 1, 1983:

> The expectation of some Western observers that the new Soviet administration would reduce Soviet military presence in Afghanistan and relinquish its pressure on Poland was no more than wishful thinking not based on any real understanding of the new Soviet power structure.[19]

6

Delusions before the Disaster

> ***"The only thing that would change the behaviour of the Soviet Union would be the ascension to power of a noncommunist leader."[1]***
>
> Aleksandr Solzhenitsyn
> Nobel Laureate
> June 21, 1983

Fred Smith, administrative assistant to Congressman Lawrence McDonald, eyed with cold rage and contempt the Central Intelligence Agency official who had come to his office to brief him after the initial confirmation that KAL 007 had been shot down by the Soviets. Smith, with his son, had been at the Pentagon's National Command Center until past midnight on September 1, 1983. The duty officer, a U.S. naval commodore, had broken the news that "it looks like a shootdown, but the evidence is fragmentary."

Smith asked the CIA official, before a group of reporters, if he could tell him any more than what he had learned at the Pentagon. When the CIA agent said no, Smith exploded: "Then get the hell out of this office. I don't know what you guys are going to do about this disaster for détente, but we are sure as hell going to get even."[2]

Smith's angry outburst might be written off as emotional, caused by his grief at losing a man he admired, believed in, and had worked for, for nine years. However, from the time in November 1982 when Andropov succeeded Brezhnev to the day the KAL 007 was shot down, delusion and wishful thinking had dominated America and other Western countries when it came to the change in the Kremlin.

James Reston of the *New York Times*, for example, wrote on January 8, 1983, that a Reagan-Andropov summit might dispel mutual suspicions if it were held in some neutral Scandinavian country or "maybe even somewhere at sea, where they might be able to talk privately without newspaper reporters and television cameras."[3]

Joseph Kraft, in his *Washington Post* column, called on President Reagan to break through his own bureaucracy and make a general speech asserting the priority of peace. Kraft wrote:

> Andropov plainly wants a deal. He came to the leadership largely through the backing of the Soviet military. They have extracted promises of more vigorous defense efforts if arms control accords are not reached with the United States.[4]

Three days before the Kraft column was published, it was reported that the U.S. Arms Control and Disarmament Agency was concerned that the Kremlin was electronically hiding data transmitted by tested Soviet missiles, thereby seeking to prevent U.S. monitoring verification as required by past arms control agreements.[5] On February 8, 1983, the Soviets tested an improved intercontinental ballistic missile that exceeded the limits of the 1979 SALT II Treaty, which had been agreed to by both the United States and the USSR, but not ratified by the U.S. Senate. The Reagan administration had pledged to abide by the treaty.[6] As we shall discover, the violation of arms control agreements may have played a role in the Soviets' shooting of KAL 007.

During the spring of 1983 the Reagan administration was under pressure to set up a summit meeting with Andropov and to push for arms control agreements. In both Europe and the United States, the mushrooming "nuclear freeze" movement was applying pressure. The KGB at the same time was making a concerted effort to infiltrate and manipulate antiwar groups in Western Europe and Scandinavia. "The degree of Soviet success thus far has been great," observed KGB defector Stanislav Levchenko in July 1983. "The buildup of criticism on nuclear weapons by these groups has gone basically in only one direction—against NATO."[7]

According to Levchenko, a KGB agent for nine years until his defection in 1979, Andropov transformed the KGB into a highly professional organization with special emphasis on a propaganda war against the West. Also, Andropov took a particular interest in overseas operations and would meet new foreign agents personally. "It is absolutely certain," he added, "that he was directly concerned in many aspects of the planning of big overseas operations. He has a liking for that kind of thing."[8]

Former U.S. ambassador to Moscow, Malcolm Toon, labeled as "Soviet disinformation" a flood of news media stories, after Andropov succeeded Brezhnev, that the former KGB chief was a "closet liberal" who liked Western clothes, music, art, and books. Toon also contrasted these assertions in the Western news media to the Soviet record of using poison gas in Afghanistan and Southeast Asia and the clear implication that the KGB was behind the attempt on the life of Pope John Paul II. "If the KGB was involved," Toon wrote, "then its chief at the time—Yuri Andropov—must have had specific knowledge of and given his personal approval to the assassination attempt."[9]

Despite Andropov's clear record of ruthlessness and a growing volume of evidence of Soviet arms control violations, a drive continued in the United States to force a Reagan-Andropov summit. Senator Charles Percy (R–Ill.), chairman of the Senate Foreign Relations Committee, openly called for such a summit on March 16, 1983, while urging the Reagan administration to compromise its own proposals for arms control in Europe. It was "unconscionable," Percy insisted, "that there were no plans for such a meeting," adding that there were "stonewallers" within the Reagan administration who were opposed to constructive relations with the Soviet Union.[10] Six days before Percy was calling for a summit, the Soviets had denounced a recent speech by the president, saying that Washington "can think only in terms of confrontation and bellicose, lunatic anticommunism."[11]

Mr. Reagan had said that the Soviets were willing to use their military power to extend their influence worldwide, producing a newly published Pentagon study to back up his assertion. Immediately Sen. Edward Kennedy (D–Mass.) called the study "classic scaremongering," and Sen. Gary Hart (D–Colo.) suggested that the report "makes it sound as if the Russians are just over the horizon."[12]

Thus, the president was pressured by liberal members of both major political parties, by the media, and by his political advisors to deal with the peace issue. Polls reported that his popularity was slipping among women on the peace issue. He was also being pressured from other quarters by conservative senators to act on an April 20, 1983, report that asserted the Soviets had violated terms of the 1979 strategic arms accord.[13]

A few days after that report was made public, Yuri Andropov launched a peace offensive ploy by publicly answering a letter from Samantha Smith of Manchester, Maine, after the ten-year-old had written to the Soviet premier asking why he wanted to conquer the world, or at least the United States. "Yes, Samantha," Andropov replied in a widely publicized reply, "we in the Soviet Union are endeavoring and doing everything so there will be no war between our two countries, so that there will be no war at all on Earth."[14] The fifth-grader was later invited to the Soviet Union in one of the most sophisticated media events ever staged by any Kremlin leader.

The first indication that the pressures from both home and abroad for a Reagan-Andropov meeting were yielding results was at the Western allied summit meeting in late May 1983. Interviewed by eight correspondents, President Reagan predicted that Soviet-American relations would improve because of support [unspecified] he had received for his arms control policy.[15]

Five days before, however, a high-level White House investigation team revealed substantial evidence of Soviet violations of arms control agreements.[16] Earlier there had been a discussion about making specific violations public. "A sincere interest in arms control," editorialized the *New York Times*, "would lead the Administration to explore these problems fully with the Russians first before it contemplates provocative public denunciations."[17]

On June 1, 1983, James Reston of the *New York Times* revealed that allied leaders had pressured President Reagan for an early Andropov meeting. Reston reported that the president had suggested any such meeting be held in 1984, the upcoming election year. The columnist added, "not an ideal time for objective discussion."[18]

By June the State Department was reported to be urging Secretary of State George Shultz to consider undertaking a diplomatic mission to Moscow in the summer to discuss a broad range of issues, including a summit.[19] On June 16 Reagan administration officials were reported by the *New York Times* to have discovered a warming trend in relations with Russia and "these officials think they have heard faint stirrings of interest in movement from Moscow."[20] Soviet Foreign Minister Andrei Gromyko's stirrings were apparently stillborn, for he insisted in a statement on June 21 that only a change in American policies and attitudes could open the way to a summit meeting.[21]

In mid-July, the rebuff forgotten, nameless senior U.S. officials believed they again saw a series of small but unspecified developments from the Soviet side on arms control. But conservatives in Congress were still pressing the White House to make public Soviet violations of past arms control agreements.[22]

Edward N. Luttwak saw as delusions all the "warming trends" and "faint stirrings" of Soviet interest. The Georgetown University Soviet scholar warned:

> Détente between the United States and the Soviet Union was made of arms control talks, summit meetings, and expectations of a costless peace. We already have the arms control talks, and now pressures are mounting for a summit meeting. It would be a grave mistake to give in to that allure—to take the easy path of competing with Moscow's public diplomacy rather than confronting its military strategy.[23]

Unmoved by such arguments, the Reagan administration made it known in early August that it was ready to conclude a new grain deal with the Soviets. *New York Times* columnist Tom Wicker had proposed that Reagan seize the initiative—

and the 1984 election—by declaring a six-month moratorium on nuclear testing and challenge Moscow to do the same. Wicker wrote:

> It's a good time because the new grain agreement, compromise in the East-West negotiations at Madrid, and some useful give-and-take in the START talks [Reagan administration's Strategic Arms Limitations Talks], lead administration officials to believe the long deterioration of Soviet-American relations has been halted.[24]

In the apparent belief that this was so, on August 20, 1983, the Reagan administration lifted the embargo to the Soviets of pipe-laying equipment for use in a mammoth trans-Siberian natural gas pipeline, while insisting that it planned to retain control of high technology items transferred to the Soviets.[25] Political columnists Rowland Evans and Robert Novak labeled the lifting of the pipeline ban a victory for those in the State Department and White House who had been working so long to convert President Reagan's hard-line anti-Soviet policy and to promote a summit, with 1984 looming large.[26]

Two days later U.S. Agriculture Secretary John R. Block concluded a grain deal with the Soviets, three and a half years after the Carter administration had embargoed grain sales because of the Soviet invasion of Afghanistan. The Reagan administration's agreement included a pledge that the United States would not impose such embargoes in the future! Secretary Block called the agreement "an early building block in the effort to build a more stable and constructive relationship" between the Soviets and the United States. President Carter's former national security advisor, Zbigniew Brzezinski, was of a different mind. "What is truly distasteful," he said, "is Secretary Block crawling on his knees to Moscow."[27]

With the grain deal and the lifting of the ban on pipeline technology in his pocket—in other words, with a pipeline to be built with slave labor and grain to feed the Soviet military—Andropov told a group of liberal Democratic senators in a Moscow meeting that a summit with Mr. Reagan would be "meaningless," since the administration has passed up so many Soviet peace proposals! Senator Dennis DeConcini (D–Ariz.) reported that he was "taken aback" by the sharpness of Andropov's tone.[28]

The world would be "taken aback" at the Soviets' shooting down, a week after the conclusion of the Soviet grain deal, of commercial flight KAL 007 and the destruction of 269 lives.

In June 1983 Aleksandr Solzhenitsyn was asked whether the change of leadership in the Kremlin from Brezhnev to Andropov would bring about any change for the better. He replied:

> Not in the least. I am constantly struck by the ignorance of so-called experts on Soviet matters, who seem to think that a change of one face, or many faces, in the Kremlin, can possibly affect how Soviet communism will act. This is a naive view of communism, not based on observation. The only thing that would change the behaviour of the Soviet Union would be the ascension to power of a noncommunist leader.[29]

In the same month Congressman McDonald, in a series of articles detailing the background of Andropov and the activities of the KGB, observed that the new Soviet leader was a throwback to the past since he had become not only the general secretary of the Communist party but also had assumed the post of chairman of the Military Council. On June 7, 1983, McDonald told the House:

> In this position Andropov is commander-in-chief of all the armed forces of the Soviet Union and thus takes direct responsibility for any deployment of troops or military decision of any shape or form. This highly centralized command and control system reaching the very top of the Soviet governmental and party structure reminds one of the situation during World War II when Stalin also had both positions.[30]

7

Defeat without War

"Without nuclear weapons, the Soviet Union would not be a superpower." [1]

Professor Seweryn Bialer
Columbia University
May 5, 1983

The Kamchatka Peninsula is eight hundred miles long and is one of the most remote and barren parts of the Soviet Union, in Soviet Asia off the Aleutian Islands of Alaska. On the night KAL 007 overflew the southern tip of Kamchatka Peninsula, the Soviets were planning to test-fire a supersecret intercontinental ballistic missile that would have impacted on Kamchatka. Alerted to this test-firing, the United States dispatched an RC-135 reconnaissance aircraft from Sheyma Island airbase at the tip of the Aleutian Islands to monitor the missile test electronically. The purpose was to discover whether the test was in compliance with past arms control agreements. This is why "Cobra Ball," the RC-135 aircraft's code name, flew within seventy-five miles of KAL 007, traveling in an opposite direction on its way back to its home base after it became clear that the test-firing had been canceled.[2]

Three days after the destruction of the Korean airliner and in the middle of a global furor over the midair mass murder, the Soviets did in fact test-fire a three-stage solid-fueled missile in the Kamchatka area, but it failed. The United States maintained that the missile test was in violation of previous arms control agreements with the Soviets.[3]

No doubt exists in the minds of arms control experts that during most of 1983 the Soviets repeatedly violated past arms control accords. It was reported, for example, in the immediate aftermath of KAL 007's destruction, that U.S. reconnaissance satellites revealed the elaborate lengths to which the Soviets had gone to conceal work on a secret new missile system on the Kamchatka Peninsula and Sakhalin Island, sometimes using sliding roofs and tarpaulins.[4]

United States Sen. James McClure (R–Idaho) took to the Senate floor on September 15, 1983, maintaining that public attempts by the Reagan administration and the national news media to separate the airliner tragedy from arms reduction talks were challenged by the facts. McClure pointed out that in the wake of KAL 007 the Kremlin leaders "have claimed the right to shoot down U.S. military reconnaissance aircraft which are engaged in the national technical means of verification of SALT treaties."[5]

As early as March 1983 McClure and other conservative lawmakers in the Senate vainly sought to warn the Reagan administration and the nation that the Soviets were systematically violating agreed-upon arms control accords. Initially, President Reagan did not dispute McClure's assertions. When the Soviets had test-fired a new PL-5 missile in February (similar to the one slated for testing the night KAL 007 overflew Kamchatka), the president himself had said: "This last one comes closest to indicating that it is a violation."[6] Later, Mr. Reagan said there were grounds to question Soviet compliance with arms control agreements; and administration officials confirmed that analysts have "conclusive evidence" that the PL-5 test "goes far beyond treaty definitions of allowable modifications."[7]

In a detailed letter to President Reagan dated March 23, 1983, and on the Senate floor the same day, Senator McClure made his case against the Soviets. "President Reagan's own recent accusations," McClure told the Senate, "that the Soviets are in violation of SALT II is thus conclusively supported by strong, unclassified evidence and consistent, reasonable interpretations of the relevant SALT II provisions by two administrations."[8]

Between McClure's public challenge in March and the KAL 007 incident, the administration demonstrated a reluctance to call the Kremlin to account for its documented arms treaty violations. On May 30, 1983, for example, the Soviets illegally tested for a third time their PL-5 missile, a test the United States for the first time was able to verify as a violation by a special radar ship stationed off the coast of Siberia in the Bering Sea.

Columnists Rowland Evans and Robert Novak reported that the radar ship had caught the Soviets red-handed. But domestic political considerations had by June 1983 intruded, presidential advisors fearing that Mr. Reagan's openly calling the Soviets to account would reinforce the president's Nuclear Napoleon image—an image bestowed on him by the Democrats. On June 29 Evans and Novak wrote:

> Both President Reagan's political advisors and the State Dept., taking different routes, have arrived at the same destination: agreement that now is the time for some U.S.-Soviet *reconciliation* [author's emphasis], specifically a new arms control agreement. Just at this point, ironically, the President will soon have on his desk evidence of Soviet cheating so blatant that it could produce an outcry menacing even U.S. adherence to SALT II.[9]

The administration's drive for a United States-Soviet summit and arms control agreement in July and August of 1983 was in sharp contrast to three separate developments.

First, on August 5, 1983, a study released by the Washington-based Heritage Foundation revealed that the Soviets had violated the 1972 Antiballistic Missile (ABM) Treaty.

> Recent photographs taken by a U.S. surveillance satellite on a routine sweep of the eastern Soviet Union reveal the construction of an immense radar system deep inside the Soviet Union north of the Mongolian border. This new radar system is targeted toward Alaska. . . . Construction of the new radar complex with a transmitter building almost five hundred feet long and three hundred feet wide violates the ABM Treaty.[10]

As early as April 1983, moreover, former State Department officials familiar with arms control issues were openly warning the Reagan administration that the Soviets, since signing the 1972 ABM treaty, had been using arms control agreements for a decade as a smokescreen to invest large sums to improve their ABM defense system.

Former State Department official Seymour Weiss wrote:

> Whether Soviet efforts are in violation of the treaty is a matter of conjecture. Many experts believe they are; others that the Soviets have simply taken advantage of the ambiguities of its terms. What is not in dispute is that the once sizable U.S. lead in ABM technology has dwindled.[11]

The second development was the admission by Arms Control and Disarmament Agency director Kenneth L. Adelman, prior to the KAL 007 shootdown, that when it came to SALT I and SALT II the Soviets "have taken these treaties to the outer edge of permissable behavior and gone beyond them."[12] Perhaps even more revealing was the disclosure on August 23, 1983, by the Reagan administration that the Carter administration concealed the discovery of Soviet attempts to "impede U.S. verification capabilities" in order to gain 1979 Congressional approval of SALT II! The disclosure was contained in a confidential Reagan administration report leaked to the press.[13]

The third development took place four days before the Soviets shot KAL 007 out of the skies. On August 27, 1983, a senior administration official was quoted by the *New York Times* as expressing "disappointment" at a proposal by Yuri Andropov, offering to dismantle missiles in Europe only if the United States abandoned its December 1983 planned deployment of Pershing medium-range missiles in Europe. Significantly, the Andropov proposal reserved the right to move any dismantled missiles from Europe to Soviet Asia, intimidating China, Japan, South Korea, or any U.S. allies in that region.[14]

On May 9, 1983, U.S. officials had confirmed that the Soviet Union was building missile bases in Siberia to double the number of its medium-range missiles aimed at the Near and Far East.[15] It was from this region, four days after the Andropov proposal, that the Soviets were preparing to test-fire their secret new PL-5 when KAL 007 overflew the Kamchatka Peninsula.

On the same day that the Soviets finally did carry out the test in violation of existing arms agreements, chief U.S. arms negotiator Paul Nitze stood in the Rose Garden with President Reagan and told the nation that, despite the midair murder, arms talks would continue. On September 3 Ambassador Nitze told reporters:

> We are all deeply concerned about irresponsible Soviet action which led to the death of 269 persons, including over 50 Americans, aboard the Korean Air jet airliner. We must nevertheless continue our efforts to reduce the threat of nuclear conflict through negotiated, verifiable agreements.[16]

The *Wall Street Journal* pointed out ten days after Nitze's statement that the Soviets' conduct in the wake of KAL 007 involved a flood of lies, denials, and eventual admission of shooting at the airliner, coupled with charges that the airliner was a spy plane, and a brutal warning that if any other aircraft violated Soviet airspace it would suffer the same fate. The *Journal* went on to point out that such conduct is consistent with Soviet behavior over arms talks, using them to inhibit U.S. weapons' development while leaving the Kremlin unfettered and free to violate past agreements. The *Journal* editorial added:

> After seeing what it is like to talk to the Soviets about downed airliners, finally, consider what it is like to talk to them about their compliance with arms treaties. Space satellite photos recently revealed a new large phased-array radar near the Siberian village of Abalakovo, capable of ABM battle management and near Soviet intercontinental missile fields. This is clearly a violation of specific and detailed provisions of the SALT I Treaty. When we protested, the Soviets blandly replied that it is a space-tracking station. From the radar's orientation, our analysts consider this implausible. The Soviet response is a lie like those told in the airliner episode, but what are we to do?[17]

The *Journal* suggested suspending the Geneva arms reduction talks, which the Reagan administration has refused to do. Instead, in conjunction with Republican Senate leaders, the administration orchestrated a toothless Senate resolution condemning Soviet action while successfully turning back a series of punitive measures that were supported by a group of Senate conservatives led by Sen. Jesse Helms (R–N.C.). Noting that the Soviet PL–5 missile test was to have impacted on

Kamchatka in violation of past arms agreements, Helms argued that, like any criminal, the Soviets killed unwitting witnesses to their violations or intended violations.

Helms told the Senate:

> In my judgment the shooting down of KAL 007 was an attempt to cover up Soviet arms violations. There are those that say that the KAL 007 flight had nothing to do with arms control and that we should redouble our efforts at Geneva. The senator from North Carolina says that we should redouble our efforts at Geneva because—I repeat—the Soviet shooting down of KAL 007 is intimately related to Soviet practice in attempting to hide SALT violations. Unless the Soviets accept responsibility for their actions, we cannot be confident that the START negotiations will result in real reduction.[18]

Five days later President Reagan told allied leaders that the United States was prepared to be more flexible in seeking an accord with the Soviet Union on limiting medium-range missiles despite the rift over the South Korean airliner.[19]

Later, at the United Nations, on September 26, 1983, the president disclosed that while the Soviets had violated arms agreements in the past, he would agree to limiting the number of missiles the United States would deploy in Europe—a partial concession to the Soviets.[20] The next day Soviet Foreign Minister Andrei Gromyko called the Reagan proposal "lopsided" and an attempt to block any agreement.[21] This was followed by a bitter blast from Andropov himself, accusing the United States of pursuing a "militarist course" that raised the threat of nuclear war.[22]

Thus, the strategy of seeking not to offend the Soviets, contrary to Mr. Reagan's instincts as a hard-liner with Russia, had ended in failure. Columnists Rowland Evans and Robert Novak reported that every time the president wanted to go public on Soviet arms violations, his advisors were against it. "His diplomatic advisors," they wrote on the day of Mr. Reagan's U.N. speech, "feared that would ruin U.S.-Soviet arms control talks, and his political advisors worried about restoration of Reagan's Attila the Hun image."[23]

Perhaps Congressman Lawrence McDonald discerned the

central flaw in the presidency of Ronald Reagan fourteen months before the downing of KAL 007. He told this author in an interview in June 1982:

> The fundamental problem with President Reagan is that he is not a hardball player when it comes to politics. He's very persuasive. But I have come to the conclusion he's the type of person who likes to sit around a table for a friendly chat, work out any disagreements and try and resolve them within an hour and a half, work out an agreement, reach a general compromise, everybody smiles, shakes hands, and then he says, let's go to dinner.[24]

During his nine years in the House of Representatives, McDonald opposed every single effort at arms reduction, believing that it would benefit the Soviets and their military strength more than the United States. He was convinced that far from wanting disarmament, the Soviets needed nuclear weapons more than the United States as the one certain means to terrorize the world into submission without war. Columbia University professor of political science, Dr. Seweryn Bialer, made much the same point in May 1983.

> In the great debate over nuclear weapons and how to control them, one key fact gets little attention. The Soviet Union depends on its nuclear arsenal not only to protect itself and to threaten others, but for its status as a great power. Without nuclear weapons the Soviet Union would not be a superpower.[25]

8

Andropov's Imperial Pacific

"Korea is geographically a dagger pointed at the heart of Japan."[1]

Sen. Jake Garn (R–Utah)
February 1976

Nine days before the KAL 007 tragedy, Japanese Defense Force radar confirmed on August 22 the arrival for the first time of a dozen modern Soviet MiG-23 fighters landing on Etorofu Island, 150 miles east of Japan's northernmost island of Hokkaido. With a cruising speed 2.3 times the speed of sound and a range of 600 miles, the MiG-23s could easily strike at Japanese airfields in northern Japan. One Japanese official said that the MiG-23s were an apparent effort to counter U.S. plans to deploy F-16 fighters in late 1984 in the northern part of Japan.[2]

Since 1978 the Kremlin has been steadily building up its military muscle on the four islands north of Hokkaido, which were seized by the Soviets at the end of World War II, in a clear effort to terrorize the Tokyo government into loosening its ties with the United States and to restrain its growing relationship with South Korea. This campaign took on a particularly brutal tone after Yuri Andropov succeeded Leonid Brezhnev in November 1982. For example, Soviet press criticism regarding Japan-U.S. relations went so far as to imply that Japan might become a target of Soviet nuclear attack. On January 25, 1983, Japan's deputy minister for foreign affairs, Toshijiro Nakajima, delivered in person an unusually blunt diplomatic protest to the Soviet ambassador to Tokyo, Vladimir Pavlov.

Toshijiro told Pavlov:

> The recent rapid Soviet military buildup in the Far East is creating a major instability for the peace and stability [sic] of this region. Recently, Soviet military aircraft not formerly stationed on our northern territories [the Kurile Islands, occupied by the Soviet Union] have flown into these territories. It is most regrettable that, despite the repeated protests of the government of Japan, the Soviet Union has continued its military buildup there.[3]

The Japanese deputy minister for foreign affairs rejected Soviet propaganda blasts against Japan's prime minister's open efforts of closer military cooperation with the United States and South Korea. (They were particularly critical of an official state visit to South Korea by Prime Minister Yasuhiro Nakasone.) "We cannot but view," Nakajima added, "the suggestions that Japan might be targeted for Soviet nuclear attack as a wanton attempt to fan anxiety among the Japanese people."[4]

Japan is the only modern nation to have experienced firsthand the effects of nuclear weapons; and the Soviets, during the months prior to the KAL 007 tragedy, kept up a steady and brutal propaganda campaign of terror. "Facts demonstrate," intoned a Tass news analyst on August 2, 1983, a month before KAL 007, "that forgetting the lessons of Hiroshima and Nagasaki, Japan's authorities throw the door wide open to these [U.S. nuclear] forces, dooming the country to the role of a hostage in the Pentagon's military adventures."[5]

Robert Keatley, Hong Kong-based Asia correspondent for the *Wall Street Journal*, took note in the early months of 1983 of the brutal and bullying tone Moscow directed toward Tokyo, terming the campaign, so soon after Andropov took power, as crude, clumsy, and unexpected.

> None of this is what experts on the Soviet Union expected when former KGB director Andropov succeeded Mr. Brezhnev as party general secretary. That's partly because they consider Mr. Andropov an intellectual by Politburo standards. Through his secret police network, he is said to grasp Soviet reality, while his many contacts with foreigners are said to have given him much knowledge of the world. Mr. Andropov has not only been

abroad, he's said to have an inquisitive mind and to be a serious reader, rare qualities in the Kremlin.[6]

Such a view was dealt a setback on May 7, 1983, when it was revealed that the Kremlin under Andropov was preparing to double its SS-20 medium-range missile force targeted on Asia in general, and on Japan in particular. Arms reduction advocates in the West were also dealt a setback when Andropov made it clear that he would maintain the Moscow line that any Soviet missiles dismantled and withdrawn in Europe would be shifted to Asia as part of its growing military buildup to achieve military dominance in the Pacific.[7]

Soviet ambitions to become a Pacific power involve beefing up not only its nuclear and conventional air and land forces, but an enormous expansion of its sea power. In March 1983 U.S. naval experts reported that the Soviets had made Cam Ranh Bay, South Vietnam, a full-time base for the operation of their Pacific fleet. The Soviets fell heir to the anchorage facilities, built by the United States, much in the same way their North Vietnamese allies fell heir to billions in captured U.S. arms. As many as twenty Soviet ships were photographed anchored at Cam Ranh Bay, the largest number of ships since the Soviets began using the facility in 1980. The Soviets were also reported to be using the huge naval complex for launching air reconnaissance aircraft to patrol the South China Sea and the Indian Ocean. They have also built an electronic intelligence complex at Cam Ranh Bay to monitor U.S. communications to Clark Air Force base and Subic Bay in the Philippines.[8]

In the same month of the stepped-up use of Cam Ranh Bay by the Soviets, veteran naval officers were warning Congress and the country that the Soviets were placing major emphasis on seapower in the Pacific. For example, Petropavlovsk, the Soviet Union's ice-free port on Kamchatka Peninsula which KAL 007 overflew, is the home port for about one-fourth of the Soviet Pacific fleet. Retired Rear Admiral Robert J. Hanks, former director of strategic plans and policy for the U.S. Navy, maintained that the Pacific fleet is the largest of the four Soviet fleets worldwide.

Rear Admiral John L. Butts, director of naval intelligence, warned Congress in March 1983, "The steady, year-after-year expansion of Soviet Navy capabilities shows no signs of slowing down. The Soviet Navy now has enough modern ships and aircraft to provide a naval presence around the world, and they are doing it on a day-to-day basis."[9]

The Soviets' control of Cam Ranh Bay puts the Kremlin in the position, with its Pacific fleet, to threaten the U.S. lifeline to the Indian Ocean, should Washington be forced to send a Rapid Deployment Force to the Persian Gulf to defend the oil fields in the region. New York University professor Albert L. Weeks maintains that the Soviets, eight years after U.S. involvement in Vietnam ended, has at least twenty countries in Africa and Southeast Asia on which they could rely in a crisis over raw materials, such as oil in the Persian Gulf or precious and strategic minerals in southern Africa. He maintains further that one critical aspect of Andropov's "Grand Strategy" was to "indict or deny shipments of strategic minerals, etc., to the capitalist industrialized West."[10]

In the aftermath of American withdrawal from South Vietnam, when the United States was planning the withdrawal of its forces from South Korea, the Soviets from 1976 to 1983 sought to supplant the United States as a Pacific power by the enormous buildup of their military might in Asia. As early as 1976 Sen. Jake Garn (R–Utah) pointed out that the Soviets were intent on becoming a Pacific power, while their proxy, North Korea, was making extensive military preparations.

> The Republic of Korea is a key part of the strategically important Northeast Asia area of the world. It has often been pointed out that Korea is geographically a dagger pointed at the heart of Japan. A Korea in the hands of a hostile power would be a serious menace to the security and independence of the strongest non-Communist nation in Asia. It is on and around the Korean peninsula that the interests of four great powers in Asia [the Soviet Union, China, Japan, and the United States] interconnect. It has been the great misfortune of the Korean people to see their part of the world fought over by great powers three times in this century [the Russo–Japanese War in 1905; the Second World War in 1945; and the Korean conflict in 1950].[11]

In the nine months after Andropov came to power, the steady Soviet buildup of military power in Asia was combined with threats against South Korea and Japan and an internal espionage campaign. Radio Moscow launched a particularly bitter denunciation of Japanese Prime Minister Nakasone's plans to beef up Japanese defenses in the face of the growing Soviet military buildup. Soviet Lt. Gen. Dmitry Antonovich Volkogonov warned over Radio Moscow that "recent words and specific deeds are dangerous for Japan and the world," adding, "The USSR cannot look on with indifference at the events which are developing."[12]

In the wake of the KAL 007 atrocity, the *Wall Street Journal* was one of the few U.S. publications to connect the downing with the previous Soviet campaign of trying to terrorize the Japanese into submission by threats. The *Journal* noted in an editorial:

> The Soviets have directed their threats at Japan, brazenly promising dire things if the Japanese proceed with their plans to build up their defense forces. Shooting down a South Korean airliner in the airspace that divides Soviet and Japanese territory would perhaps be some politburo bigwig's idea of how to drive that message home.[13]

Congressman Lawrence McDonald was on his way to Seoul to attend a conference marking the thirtieth anniversary of the U.S.-Korean Mutual Defense Treaty. In the speech he had planned to deliver, McDonald said that the economic prosperity of the South compared to the North since the end of the Korean conflict in July 1953 made it possible for Seoul to be selected as the site for the 1988 Olympic Games. "A fact that must cause that bleak regime in the North to grind their teeth with hatred," the Georgia Democrat observed, adding:

> The military and security forces of the Republic of Korea have captured and broken scores of Communist espionage, sabotage, and terrorist cells. Some have been infiltrated through the secret tunnels, others by land or boat. They have broken up subversive rings of students and intellectuals acting on the orders of the Communist North to carry out "active measures" [assassination] of subversion and propaganda to discredit and undermine South Korean leaders and institutions.[14]

A month after McDonald's death, a bomb blast in Rangoon, Burma, went off during an official state visit by the president of South Korea, Chun Doo-Hwan, killing sixteen, including the foreign minister and several cabinet members responsible for the economic miracle in South Korea. As with the KAL 007 shooting, the Reagan administration urged that Seoul exercise "restraint." The *New York Times* quoted senior U.S. officials to the effect that "South Korea should not retaliate with force if it turned out that North Korea was responsible for the assassination of sixteen South Koreans, including four cabinet ministers"![15]

It was not reported what measures, if any, the administration thought were appropriate.

9

Motives for Mass Murder

***"This is a watershed event. For a 'cheap' price the Soviet Union has demonstrated its credible military power."*[1]**

Dr. Arnold Beichman
coauthor *Andropov: A Political Biography; New Challenge to the West*
September 5, 1983

"Sometime this year," wrote Dr. Arnold Beichman three weeks before KAL 007 was shot down, "certainly before December [1983], when the cruise and Pershing II missiles are due for deployment in Western Europe, the Soviet Union (with or without an ailing Yuri Andropov) will do something that will precipitate a confrontation between the two superpowers. The confrontation will be one for which American public opinion will be ill-prepared."[2]

In the immediate aftermath of the midair massacre, it was widely suggested in the Western news media that the downing of the airliner was the result of a trigger-happy Soviet pilot or a breakdown in Soviet military command structure or even an outgrowth of Soviet paranoid preoccupation with territorial security. Almost no effort was made by reporters or analysts in the media to explain the act in terms of cold-blooded premeditation with a specific political objective in mind.

United States Soviet strategy analysts William C. Green and David B. Rivkin, Jr., experts on Kremlin leadership psychology, pointed out that the widely broadcast view in the West that destruction of KAL 007 was an outgrowth of "paranoia" amounts to extending to the Kremlin the American legal doctrine of the insanity defense, which mitigates or absolves Soviet guilt and responsibility.

> This view of Soviet behavior is not substantiated by available evidence. While the Soviet press routinely assigns the most malign intentions to the West, actual Soviet behavior shows a realistic assessment of Western capabilities and will. Therefore, the more tenable view is that the Soviet Union is a dispassionate, pragmatic, and cold-blooded superpower which does not shirk at any action which would serve its political goals. The shooting down of KAL Flight 007 should be no exception to this pattern.[3]

Drew Middleton, military analyst for the *New York Times,* came closer than anyone else in the media to assigning a specific cold blooded motive to the Soviets. Noting they had been building up their military power for twenty-seven months in the Kuril Islands chain, specifically on Sakhalin Island and the Kamchatka Peninsula, Middleton concluded that Soviet sensitivity over penetration of their radar defense guarding the La Perouse Strait south of Sakhalin—in war, a direct route for deployment of the Soviet Pacific fleet into the northwest Pacific—was behind the order to destroy KAL 007. Middleton wrote:

> American and NATO intelligence analysts rejected the idea that the shooting down of the Boeing airliner could have been an impulsive act of a Soviet pilot. They cited the rigid, centralized command system of Soviet forces and said an attack on a civilian airliner, even when it was in Soviet airspace, could not have been carried out without the approval of higher headquarters.[4]

The evidence before, during, and after the destruction of KAL 007 reveals a constant Soviet preoccupation with matters military, specifically in Asia and around Sakhalin Island and the Kamchatka Peninsula. Professor Albert Weeks of New York University concluded that the Soviets' destruction of KAL 007 was no accident but was deliberate and premeditated. Dr. Weeks pointed out that in a June 1983 Communist party central committee meeting directives were issued from the top Soviet civilian and military leadership that placed its armed forces in key defense zones near Soviet borders in a state of virtual alert. Weeks cited a June 27, 1983, speech by Soviet Defense Minister Dmitri F. Ustinov, a member of the ruling politburo: the marshal called on new Soviet officers and their senior colleagues throughout the military "to do everything necessary in order to always make our arms piercing and terrifying for our enemies."[5]

Up to a few days before the destruction of KAL 007, the Soviet press was filled with such statements, suggesting that the top Kremlin leadership was planning some major premeditated operation without revealing its details. Dr. Weeks observed:

> At the June 1983 Central Committee plenum General Secretary Yuri Andropov and the once-intended heir to the late General Secretary Leonid Brezhnev, Konstantin Chernenko, both sounded the note of extreme preparedness "because the international situation lately has seriously worsened."[6]

During Andropov's fifteen years as head of the Soviet KGB, one of his major functions was to police the political loyalty of the Soviet military, which now provides him with close ties. Author John Barron, in his published study of the KGB, pointed out that since 1926 it has been part of Soviet policy to use assassination and murder. Today a special branch of the KGB, which it most wishes to keep secret, exists for such purposes—the Executive Action Department, or Department V—as in *Victor.* Barron wrote:

> For this ultrasecret department is responsible for the Soviet Union's political murders, kidnappings, and sabotage—actions which, in the KGB parlance, are called "wet affairs" *(mokrie dela)* because they often entail the spilling of blood. . . . Its purpose is to give Soviet rulers the option of immobilizing Western countries through internal chaos during future international crises.[7]

In the days after the destruction of KAL 007, the Soviet news agency Tass put out the story that former President Richard Nixon was slated to attend the Seoul meeting with other U.S. representatives. But, said Tass, Nixon was told by the CIA at the last moment to cancel the trip because it knew the airliner was a spy plane. A Nixon spokesman labeled the story a fabrication.[8]

The principal organizer of the Seoul conference revealed that Mr. Nixon had been invited to attend but had turned down the invitation. Garrett N. Scalera, president of the Tokyo Institute of Policy Studies, one of the principal sponsors of the conference and its overall coordinator, is convinced that the Soviets shot the airliner down for a specific reason. He told this author:

> The Soviets saw the conference as a threat to their long-range plans to terrorize and eventually dominate free Asian nations. The purpose of the conference, as well as an earlier one in Tokyo, was to begin laying the groundwork for a northeast Asia security framework.[9]

A NATO-type military alliance for the Far East, with backing of U.S. senators and congressmen, would represent a challenge to the Soviets' plans in the Pacific. Scalera maintains, moreover, that the Soviets knew the names of those Americans who planned to attend the conference, when they were scheduled to leave Washington, and by what mode of transportation.

> Some of the toughest and most militant anti-Communists in Congress were slated to attend, and wiping out important and vocal conservative anti-Communists like Senator Jesse Helms (R–N.C.) would have set back the work we were doing to form a security framework for northeast Asia.[10]

Lilly Fediay, staff member of the Washington-based Institute of American Relations and cosponsor of the conference, said that she gave the Korean embassy in Washington a complete list of the names and airline schedules for the U.S. congressional delegation traveling to Seoul. U.S. intelligence experts maintain that the Soviets routinely monitor Telex and cable traffic from foreign embassies in Washington.[11]

The Soviets, therefore, probably knew in advance that Senator Helms, Sen. Steve Symms (R–Idaho), and U.S. Rep. Carroll Hubbard, Jr. (D–Ky.) were slated to take KAL 015 from Los Angeles to Seoul with a refueling stop in Anchorage. Several others were scheduled to take the same flight as Helms, Symms, and Hubbard: Ron Mann, associate director and presidential aide for security affairs; Dr. Donald Stims, deputy undersecretary of defense for nuclear targeting; William Scheider, Jr., undersecretary of state for security; and Robert McCormick, assistant secretary of state for security. At the last minute they decided to take a flight from New York on Sunday, August 28, which Congressman McDonald missed by only minutes.

Kathryn McDonald recalled later that in the last telephone conversation she had with her husband, she pleaded with him to cancel his trip to Seoul because she had an uneasy feeling and was concerned that McDonald was exhausted from his heavy workload."I have to go," Mrs. McDonald quoted her husband as saying. "I have given my word and as much as I would like to come home for a rest with you and the children, the South Koreans are counting on my being there."[12] McDonald stayed over in New York until Tuesday August 30 to take care of some private business, according to Mrs. McDonald, and boarded the ill-fated KAL 007 there.[13]

Bruce E. Herbert, deputy director of the Washington-based Center for International Security, does not rule out the possibility that the Soviet fighters confused KAL 015 and KAL 007 since they were only fifteen minutes apart.

> With such tempting targets [the Congressional and security passengers], it may have been difficult for the KGB to decide which aircraft to destroy. If the strongly circumstantial evidence of Soviet complicity in the attempted assassination of the Pope is correct, it hardly seems likely that they would hesitate to shoot down a civilian airliner in order to reaffirm brutally their contempt for the free world while at the same time eliminating avowed enemies of the Soviet state.[14]

The Soviet analyst maintains further that the destruction of KAL 007 is consistent with the character and cunning of the KGB. "Given the past history," adds Herbert, a former naval transport pilot with long experience, "of furious and *immediate* reaction to penetrations of their airspace, *the Soviets must have known in advance* that the airliner would enter their airspace northwest of the Aleutians, and they chose the time and place of its destruction."[15]

Herbert maintains that the two hours that KAL 007 was in Soviet airspace strongly suggests premeditation. He points out, moreover, that the belief that the Korean Airline pilot on 007 punched the wrong coordinates into the computerized Inertial Navigation System (INS) of the Boeing 747, which allegedly accounted for the navigational error without the crew's noticing it, is possible but not plausible. Produced by Litton Industries, the cassettes or "route cards" are preprogrammed to save pilots and navigators the trouble and time of punching in the navigational coordinates for latitude and longitude since the routes are so frequently flown.

Herbert stated to this author:

> The route cards are stored on board the 747 aft of the copilot's seat in a bulkhead compartment, together with the air navigational charts, Enroute Supplements, Airman's Guide, etc. They are clearly labeled and are readily available to anyone on the flight deck. While the airliner is on the ground, maintenance personnel pass freely in and out of the cockpit. Anyone wearing coveralls and Mickey Mouse ears [noise suppression headphones] around his neck would be unchallenged. It would be a matter of a moment to substitute a bogus route card for the real one.[16]

The INS aboard 747s have three separate systems in the event of failure of one or two. It was widely reported that the reason KAL 007 was so far off course was that the wrong navigational information was manually fed into all three systems. Actually, the standard procedure is to program the first two systems with the "route card" and the third by hand. "But in point of fact," Herbert stated, "knowing pilots and talking to pilots as I have and do, the reliability of the route card is so great that you routinely program all three with the cassette."[17]

A study of the INS, released shortly before the KAL 007 tragedy, found it to be almost error free; only about one flight in ten thousand strays some fifty miles off course and that is usually attributed to pilot error.[18] (KAL 007 was off course by three hundred miles.)

Until 1976 Dr. Igor Glagolev was chief of the disarmament sector of the Institute of World Economy and International Relations of the Academy of Sciences of the USSR. Escaping to the United States in 1976, where he was granted political asylum, he formed with the help of Congressman Lawrence McDonald the Association for Cooperation of Democratic Countries in an effort to mobilize world public opinion against the global aims of the Soviets, which—until 1976—he was in the position to know from firsthand knowledge. Dr. Glagolev has no doubt that the Soviet downing of KAL 007 was deliberate. He also revealed that the Soviets routinely attacked his and McDonald's organization in their publications, and KGB agents kept McDonald's activities and that of his supporters under surveillance.

It is possible, following Dr. Herbert's theory, that while KAL 015 and KAL 007 were both on the ground in Anchorage at exactly the same time for refueling, a KGB agent posing as a ground maintenance crewman might have placed a bogus route card on the wrong aircraft.

Dr. Glagolev told this author:

> Andropov and the whole leadership of the Soviet Union did what they wanted to do: show the impotence of the United States and the power of the Soviet Union. It was an act of terrorism, that they are the strongest military power and they can blackmail anyone, including the United States. More than that, they showed that even after these mass murders the U.S. government and commercial banks continue to finance the Soviet bloc. They have been organizing mass terrorist actions around the world, and the shooting down of the Korean airliner was a mass-terrorist action.[19]

Robert Reilly, one of the authors of the book *Justice and War in the Nuclear Age,* suggested that the destruction of KAL 007 must be seen in the context of Soviet military strategy.

> The uncertainty that terrorism introduces activates the mind to imagine that anything can happen. The false certainty of nego-

> tiations is a welcome substitute for the fear of the unknown that terrorism generates. Thus, the KAL incident may well be aimed at the arms-control negotiations and used, ironically, as a reinforcement to the "peace" movement in Western Europe.[20]

Dr. Arnold Beichman, who predicted beforehand that the Soviets would seek to precipitate a superpower confrontation, noted after the downing of KAL 007 that it was a crucial test for the United States and the president. According to him, Ronald Reagan played into the hands of Soviet strategy by doing nothing and insisting on arms negotiations. He added:

> This is a watershed event. For a "cheap" price the Soviet Union has demonstrated its credible military power. It's part of the Soviet strategy to take over the world. They demonstrated that international terrorism pays off when it's conducted by a government. They demonstrated that the western Pacific is their domain and made it clear to Japan what the price is if countries do things they do not like. It comes at a time when relations between Japan and the Soviet Union have been strained by the Soviets' arms buildup in the Far East and Prime Minister Yusukiro Nakasone has responded by strengthening ties to Japan's Western allies.[21]

Congressman Lawrence McDonald had spent his entire career warning against the use of terrorism as an instrument of Soviet policy, particularly the use of the threat of nuclear war by the Kremlin as a weapon to paralyze the United States and its Western allies' will to resist. He wrote in September 1977:

> If you believe, or at least accept without much mental protest, that nuclear war means total extinction, you will agree with your leaders that it must be avoided at all costs, and this prepares you nicely for the first "nuclear blackmail scenario" to come down the pike. You are ready to submit.[22]

10

An Exodus to Eternity

***"I really had a very uneasy feeling during the flight. . . . I was very glad to get to Seoul."*[1]**

U.S. Sen. Steven Symms

International air travel represents for the entire free world an important element of its worldwide economic system, particularly in terms of trade and tourism. In 1982 it was estimated that 750 million passengers traveled to their international destinations by commercial airliner.[2]

One of the key aims of Soviet terrorism is the systematic destruction of a country's free economic infrastructure and the disruption of tourism to those countries targeted for a take-over. Months before the destruction of KAL 007 by a Soviet fighter, it was widely publicized that sixty-three hundred travel agents from ninety-one countries would be meeting in South Korea for six days in late September 1983 for the Fifty-Third World Travel Congress of the American Society of Travel Agents (ASTA).[3] A majority of the ASTA delegates are expected to handle the large volume of visitors to South Korea during the 1988 International Olympic Games. The ASTA meeting was not cancelled, and the delegates passed a seven-point resolution condemning the destruction of KAL 007.

Shortly before midnight on August 30, 1983, Korean Airlines flight 007 was cleared by the control tower at New York City's Kennedy International Airport for take-off to Seoul with a refueling stop in Anchorage. The 269 people aboard were of fifteen different nationalities: 105 Koreans (including the 29 crew members of whom the youngest was only twenty years old), 61 Americans, 28 Japanese, 23 Taiwanese, 15 Filipinos, 12 residents of Hong Kong, 9 Canadians, 6 Thais, 4 Australians, 1 Swede, 1 Indian, 1 Briton, 1 Dominican, 1 Vietnamese, and 1 Malaysian.[4] The passengers represented a cross section of tourists and business and professional people. At least 115 of the KAL 007 passengers were females and 29 were children ranging in ages from one month to sixteen years. At least 12 passengers were over sixty, the oldest being eighty.[5]

The mood aboard KAL 007, according to a relative who said good-bye at the Kennedy airport departure lounge for what would be the last time, was "cheerful." One passenger had the "luck" of winning his KAL ticket in a golf tournament! Congressman Lawrence McDonald sat in an aisle seat in the first-class cabin deck of the aircraft, his mind no doubt turning to the conference in Seoul and the speech he was to deliver.[6]

"I believe that in a number of cases," one part of the speech read, "we can detect probable 'active measures' on behalf of the North Korean regime in Pyongyang [North Korea's capital] carried out by that regime's agents and on their behalf by the Soviet KGB."[7]

Sixteen days prior to the take-off of KAL 007 from New York, the South Koreans had reported sinking the sixty-ton ship *Asahi Maru,* which was disguised as a Japanese fishing boat. However, according to the South Koreans the vessel was a spy ship carrying North Korean agents for infiltrating the South. The bodies of three North Koreans were recovered, along with their personal effects and three machine guns, frogmen underwater equipment, and three pocketbooks with their contents of North Korean origin. The South Korean Defense Ministry announced that this was the first large North Korean vessel sunk by South Korean forces in its territorial waters since the end of the Korean conflict in July 1953. On August 5, 1983, another North Korean spy vessel was sunk.[8]

Thirty-three of the KAL 007 passengers were from the New York area and were Americans or naturalized citizens originally from South Korea. Several shared with Congressman McDonald a professional interest in medicine. Dr. Jong Jin

Lim, for example, was on his way to South Korea with his brother to visit their mother who was seriously ill. He was a research associate at New York City's Columbia University medical school, specializing in medical research on the lining of the cornea of the eye. Several of the American passengers were buyers on their way to Seoul to negotiate business transactions with the South Korean textile industry. Twenty-four-year-old Korean In-ho Lee, living in Jersey City, New Jersey, was on his way back to South Korea in search of a bride to bring back to his adopted America.[9]

Perhaps the most tragic and ironic of all the doomed passengers aboard KAL 007 was Billy Hong. Born in South Korea and later adopted by an American soldier, James Martin, after his parents had been killed in a North Korean air attack on their village during the Korean War, Hong had settled in Spartanburg, South Carolina, married and fathered two children. He had been a naturalized U.S. citizen for ten years and was on his way back to Seoul on business. "I am sure," observed Hong's congressman, Republican Carroll A. Campbell, "that the stories of all the innocent victims of flight 007 are just as poignant and just as tragic."[10]

The world of the terminally ill will never know what might have happened had fifty-seven-year-old Dr. Michael Truppin not taken KAL 007. Congresswoman Geraldine Ferraro (D–N.Y.) lost her family doctor of twenty-three years when Dr. Truppin perished on that flight. He had delivered three of her children. She recalled later:

> He was a leader, a pioneer, in the use of laser beam research and other advanced procedures to find a cure for cancer and sexually communicated diseases. But his greatest innovation was in having an office in which the foremost idea was to provide the best personal care for his patients. . . . he cared about his patients as people.[11]

When Dr. Lawrence McDonald was in active medical practice, prior to his election to Congress, many of his patients told similar stories. A McDonald staffer recalled that after his election to Congress, McDonald applied the same approach to politics that he had to his medical practice. Once on the way to the Atlanta airport to catch a plane for an urgent meeting in Washington, the Georgia Democrat came upon an auto accident; he stopped, gave first aid, and waited for the arrival of an ambulance. This same staffer recalls that, contrary to his public image, McDonald as a physician and as a political leader was always concerned about individuals, rather than groups.

R. D. Patrick Mahoney, McDonald's congressional office manager, recalled that Odessa Ferguson, a black cafeteria worker in the Capitol restaurant, related how her Washington church was in dire financial straits. "Larry helped bail her and the church out," he recalled to this author with a wry smile. "This was the same Larry who stood up on the House floor and opposed a national holiday for Martin Luther King, Jr. He was a very principled person, who was more interested in individuals than he was in faceless groups."[12]

We will never know whether McDonald foresaw the fate that awaited him as KAL 007 traveled from New York to Anchorage, a flight that was uneventful, with clear skies, good weather, a smooth trip west, and traveling at five hundred miles an hour—passing through several time zones. After arriving in Anchorage most of the passengers deplaned to stretch their legs. Congressman McDonald stayed aboard, perhaps asleep.

Congressman Carroll Hubbard (D–Ky.) arrived aboard KAL 015 with Sen. Jesse Helms (R–N.C.) and Sen. Steven Symms (R–Idaho). Both Hubbard and Symms had originally been scheduled to take KAL 007 out of New York.

Hubbard later recalled:

> While I was walking through the Anchorage airport, I heard them calling flight 7 for reboarding. It is very possible that had Congressman McDonald gotten off and we had found out the flight was going to arrive ten or fifteen minutes earlier than the one we were on, we might have decided to get on his flight. I don't guess we could have talked him into riding with us because 015 was almost full in first class. I will never forget seeing those people board. One of the passengers let me look through his coin operated telescope to peek at the lights of downtown Anchorage on the observation deck. I didn't know it at the time,

but I was watching doomed people for sure. I just never had an experience like that—where you were mixing and mingling with people who were soon to meet their eternity.[13]

Senator Symms and his wife, Frances, later recalled that originally the organizers of the Seoul conference had wanted everyone to travel on the same plane. Both also remember the Grenfell family in the Anchorage airport.

> He was handsome, a good looking guy. His wife was blond and very beautiful. Oh, dear God! Those two children were just adorable. Mrs. Grenfell was reading to the two children; and I remember thinking what a beautiful picture, those two little girls intently listening to their mother in the early morning, so sweet and so polite. It just breaks your heart to think of them now, both on each side of their mother—the picture of trusting innocence.[14]

When KAL 015 left Anchorage, Symms would later remember that he was overcome by a disquieting and disturbing feeling.

> I had really a very uneasy feeling during the flight. Usually when traveling on airplanes I am the easiest flyer in the world. I just don't worry about anything. But I was very restless during the flight and was very glad to get to Seoul. . . .
>
> I don't know whether it was ESP or not; but in the middle of the flight out of Anchorage, I guess it must have been about four hours out, after dinner and looking at the inflight movie, Fran was asleep and so were most of the first class passengers. I got up to go to the bathroom and on my way out I stopped and looked out the window in the rear of the plane. The moon was out and it was a clear night. I distinctly remember this crazy thought: "We are close to Russian territory; wouldn't it be just like the Soviets to look on this airplane as a fat duck flying so near they might think of shooting it down!" I then remember thinking, "What am I doing up here with Helms? It would be just like the Soviets to shoot the plane down because he's riding on it." It was just a crazy passing thought and I didn't dwell on it until later, and I was amazed at what I thought at the time.[15]

Symms calculates that at roughly the time he had this "crazy thought," KAL 007 with McDonald and 268 others was either under attack or careening toward a watery and ice-cold Pacific grave.

The Rev. Joseph C. Morecraft III, McDonald's friend and theological advisor, at the memorial service in Marietta, Georgia, made a chilling revelation to the assembled mourners.

> In the last speech Larry McDonald delivered, the Saturday before his death, he said if we are going to win this war before us, we must give ourselves selflessly and relentlessly to the advancement of the causes of God, of righteousness, and of truth until one of two things happens—we win or we are laid in our graves.[16]

11

Deadly Combination of Coincidences

***"The Soviets routinely try to lure U.S. military and intelligence aircraft into Soviet airspace so they can 'legally' shoot them down."*[1]**

Jack Anderson column
September 20, 1983

The evidence is strong but not conclusive enough to establish Kremlin premeditation in the destruction of KAL 007. The one case for clear premeditation was the 1960 order by Soviet Premier Nikita Khrushchev to shoot down a U.S. RB-47 reconnaissance aircraft in international waters off the Kola Peninsula. It was in this same area in April 1978 that Korean Airlines flight 902 experienced unprecedented navigational problems because of what the crew claimed was "an electrical shock [that] paralyzed the navigation system."[2]

As already mentioned, former South Korean Air Marshal Chang Chi Ryang maintains that KAL 902, in 1978, was the victim of a Soviet-owned electronic device that caused its navigation instruments to malfunction. Ambassador Chang also maintains that the Soviets may have used the same device to throw KAL 007 off course after it left Anchorage.[3]

Aboard every commercial airliner flying from Anchorage to destinations in the Far East are Federal Aviation Administration navigation charts, designating five separate air corridors. The one that runs closest to the Soviet's Kamchatka Peninsula is designated as R-20. KAL 007 was flying this particular corridor. On each of the FAA navigational maps, in the left-hand corner in bold type, are two warnings for aircraft crews. The first states that aircraft may be fired on without warning if they stray from their course. The second warning, just below the first, states:

> Warning: Unlisted Radio Emissions from This Area May Constitute a Navigation Hazard or Result in Border Overflight Unless Unusual Precaution Is Exercised.[4]

When KAL 007 reached Anchorage the pilot, fifty-two-year-old Kim Suh-il, reported that during the flight from New York he had experienced minor compass and radio problems. A Korean Airlines supervisor in Anchorage later stated: "Supposedly repairs had been made before the aircraft continued to Seoul."[5] Mechanics, however, reported finding nothing wrong.[6] Captain Kim and his crew turned over command of 007 to a new crew in Anchorage and remained on board as passengers. When 007 left Anchorage, therefore, it had two full crews, all experienced and responsible pilots, copilots, and flight engineers.

When forty-five-year-old Captain Chun Byung-in took command of KAL 007 in Anchorage, the passengers were in the hands of what airline officials described as a "model pilot," who the year before had received a citation for his accident-free record. Captain Chun joined KAL in 1971 after leaving the Korean Air Force and, as well as flying to the Middle East, Paris, Los Angeles, and New York, had flown the route from Anchorage to Seoul for five years.[7] Mr. Chun had also flown one of the first 747s acquired by KAL and was so highly regarded that when the president of Korea, the target of several assassination efforts, visited the United States in February 1981, Chun was chosen as the pilot. Combined with his Korean Air Force experience, he had a total flying experience of almost 20,000 hours—6,618 aboard 747 jets.[8] Captain Chun's copilot, Son Dong-hui, 47, and also a Korean Air Force alumnus who left the service in 1977 with the rank of Lieutenant Colonel, was a seasoned pilot, having logged almost 9,000 hours—3,400 hours on 747s.[9]

As soon as KAL 007 left Anchorage it was being tracked by Soviet radar. At least that is what former director of the Department of Defense intelligence, Lt. Gen. Daniel O. Graham, maintains. "The Soviets, within a few minutes after an aircraft gets to ten thousand feet," Graham told this author, "have you on their radar. They also know the schedules of the commercial airlines. They know exactly what flight it is, and they also know that no spy planes fly out of Anchorage."[10]

Colonel Samuel Dickens, retired U.S. Air Force veteran with almost his entire twenty-one-year career spent in fighter interceptors, adds that when KAL 007 initially left Anchorage, the international air traffic control system passed such information to regions that have the responsibility for monitoring such flights, including the Soviets.

> They knew just when that Korean 747 departed Anchorage. They knew the route of its flight, and the Soviets were tracking that aircraft long before it got to the Kamchatka Peninsula; and they would know this by reason of the international air traffic system that is in place. They also knew what kind of aircraft they were tracking, just as they knew where the RC-135 reconnaissance aircraft was. The difference in size of the two is so great that there is no way they could have confused the two.[11]

Sometime during the first few hours of the flight of KAL 007 a combination of coincidences took place.

First, the Inertial Navigation System (INS), which is a system of three separate computers aboard the 747 plane, was misprogrammed with the wrong flight data, so that the plane was eventually almost three hundred miles off its course. The INS is not only used in 747s but by the U.S. astronauts to guide the space shuttle to its pinpoint landings. Don Walters, a spokesman for Litton Industries, which manufactures the INS, maintained that the system is nearly "fail safe" and has a possible margin of error of two miles for every hour of aircraft flight.[12]

Within that estimated error possibility, KAL 007 should have strayed no more than ten miles off course, a safe distance and still well within the fifty-mile air corridor of R-20.

Brad Dunbar of the National Transportation Safety Board noted that in all the years the agency has investigated air crashes, "We've never had an accident investigation involving an INS failure on an airliner."[13]

The second remarkable coincidence concerned radio problems that the aircraft had encountered. Great attention was paid to the radio intercepts that documented for the world how the Soviets stalked and then shot the Korean 747 out of the skies. Almost no attention was focused on the tape-recorded conversations between the aircraft and Anchorage control and Tokyo air route control. Those transcripts, released by the Federal Aviation Agency, reveal that on at least two occasions 007 had to relay its position through KAL 015, which was behind it, because of apparent radio difficulty.

While it is not unusual for commercial aircraft to communicate with its air traffic control in this manner, it is strange that KAL 007 did it a third time through United Airlines flight 18, which was in a northern course enroute from Hong Kong to Seattle. The FAA transcript also shows that KAL 007 had difficulty communicating with Anchorage control and even missed one of its required position reports. Since the aircraft was out of range of Anchorage's radar after 165 miles, Anchorage could not verify the aircraft's position, and its reports did not show that it was off course. When the airliner did establish direct radio contact with Tokyo to report its position, about thirty minutes before it was shot down, it was apparently unknowingly well off course and being stalked by Soviet fighters.[14]

In the hours after it became clear that KAL 007 had been shot down, Congressman George Hansen (R–Idaho) and members of Congressman Lawrence McDonald's staff insisted that the plane had been lured off course electronically by the Soviets. It was not until September 20 that columnist Jack Anderson reported that the Central Intelligence Agency and the departments of State and Defense had information of Soviet efforts to scramble navigational signals along their borders and that several planes had been shot at after jamming attempts from "unlisted radio emissions." It was because of this danger over a prolonged period that the FAA quietly placed navigational maps for use on the Far Eastern polar routes near Soviet territory with their printed warnings.

Anderson wrote:

> The Soviets routinely try to lure U.S. military and intelligence aircraft into Soviet airspace so they can "legally" shoot them down. This is done by a jamming technique, called "meaconning," which confuses pilots trying to follow radio signals from the ground.[15]

The INS aboard KAL 007 did not rely on radio signals from the ground to perform its navigational task. For KAL 007 to keep within the air path designated R-20 it relied on high-frequency omni-bearing range navigational radio signals.[16] Nevertheless, it was shortly after leaving the coast of Alaska, out of range of Anchorage radar but not of Soviet radar, that the aircraft began experiencing radio trouble that lasted throughout most of the trip. No public explanation has been made in an attempt to explain this difficulty.*

It is a remarkable and deadly duo of "coincidences" that confronted KAL 007: an INS that apparently carried it off course and radio trouble once it was outside of radar contact with Anchorage. The radio trouble did not appear to have cleared up until the aircraft was close to leaving Soviet territory, had radioed Japan its supposed position (the correct one if it had been on course), and shortly before the order was given to destroy the 747 as it was about to reenter international airspace.

Officials at the National Transportation Safety Board and the Federal Aviation Agency were sent to Seoul in the imme-

*A survey of pilots and navigators conducted by *Aviation Week & Space Technology*, published a month after the destruction of KAL 007, showed that most rated INS as having extremely low error rates; but speculation centered on human errors or faulty procedures, which led the aircraft so far off course. Again, this does not account for the radio difficulties the aircraft also experienced.

diate aftermath of the airliner tragedy. On their return, officials of both agencies refused to talk publicly about any aspect of the case. William Hendrick of the NTSB, who was sent to Seoul, acknowledged that he could not recall a case involving an air accident investigation that pointed to the failure of INS. He also revealed that Korean Airlines officials had told him in Seoul that KAL 007 did not use the preprogrammed cassette or "route card" for INS, casting doubt on whether a deliberately misprogrammed cassette was placed aboard the aircraft during its refueling stop in Anchorage. Korean Airlines officials refused to talk about any aspect of the tragedy. In fact, a shroud of secrecy was drawn over the entire affair. "I don't think," Hendrick told this author, "that anyone involved in this thing wants to talk about it because there are some security aspects."[17]

We are left with the possibility that KAL 007 was hijacked at gunpoint, which would explain why the INS took the aircraft so far off course with a pilot who had worked with the system for years and had flown the polar route for five years. Experts in Seoul and Tokyo strongly leaned toward the possibility that the aircraft was hijacked, although *Newsweek* magazine revealed that two KAL sky marshals were aboard that flight, but that Captain Chun did not, as far as we know, activate the 747's hidden antihijack alert signal, which emits a powerful radio message.[18]

Jack Anderson states that a leaked top-secret CIA report indicates that in the 1978 incident the KAL passenger plane may have had a Soviet agent aboard who was instrumental in electronically disorienting the aircraft's navigational instruments.[19]

A hijacking would help explain the unprecedented failure or "misprogramming" of the INS, with its three separate programs as a doublecheck, by so experienced a pilot as Captain Chun. It would also explain the puzzling first report that KAL 007 had landed on Sakhalin Island and that everyone aboard was safe.[20] It is possible that the KAL sky marshals and the second crew regained control of the aircraft from the hijackers—and as the plane was apparently about to leave Soviet airspace it was shot out of the air. But in the absence of survivors, as there were in the 1978 incident, this theory is highly speculative, if not dubious.

Captain Kim Chang Ky, the pilot on the 1978 KAL 902 flight, later revealed that after spotting the Soviet interceptor, he reduced speed, lowered his landing gear, and flashed his navigation lights on and off—only to be hit with a missile. Captain Ky believes that Captain Chun, had he spotted the Soviet fighter, "would have taken necessary measures to insure the safety of the plane and the passengers."[21]

The evidence strongly indicates that the Soviets lured KAL 007 off course in an effort to terrorize Asian allies of the West. The planned deployment of U.S. Pershing missiles in Europe may also have played a role, with the Soviets' shooting down KAL 007 in an effort to gain cancellation of the deployment. We know from the radio intercepts published later that the Soviets systematically and cold-bloodedly destroyed KAL 007. It is, therefore, highly probable that the same ruthless premeditation they demonstrated in destroying the aircraft was also manifest in making certain that they had a target within their airspace so as to be able "legally" to justify their act of terrorism as a defense of their sovereign territory against alleged espionage. The evidence points more strongly in this direction than toward the widely-accepted version that a series of human errors in navigation, radio trouble, and even the deliberate act of KAL 007's taking a shortcut to save fuel, as has been suggested, placed the 747 deep in Soviet airspace.

We will never know what went through the minds of the 240 passengers and the 29 crew members of KAL 007 in the seconds after the Soviet missiles struck and doomed the plane to death within a short twelve- to fifteen-minute span. Senator Steve Symms (R–Idaho) believes he knew his friend Lawrence McDonald well enough to guess what went through his mind at the moment the aircraft had been hit. Senator Symms told this author:

I will always believe that Larry McDonald knew he had been shot down by the Russians. He was smart enough and his instincts were such that he looked out and saw the wing on fire; and I just feel that he probably thought to himself, "Those dirty bastards finally got me."[22]

12

A Spiraling Death Dive to the Sea

"When you come up on an aircraft of a 747's size, and I have as a fighter pilot, you are just awed by its sheer size."[1]

Col. Samuel Dickens, USAF (Ret.)
Fighter Interceptor Pilot

On February 21, 1973, a Libyan Airlines 727, enroute from Tripoli to Cairo, intruded twelve miles into Israeli airspace over the then-occupied Sinai desert and was subsequently shot down by Israeli fighter planes with the loss of 108 lives.[2] Later, at a meeting of the International Civil Aviation Organization (ICAO), the Soviet Union said it was "convinced that ICAO could not remain aloof" from such a "barbaric act." The Soviets insisted that as a specialized agency of the United Nations the ICAO was pledged to the promotion of air safety in all parts of the world, and they called for condemnation of Israel.[3]

In the aftermath of the Soviet Union's shooting down of KAL 902 in April 1978, ICAO approved new and expanded procedures on the interception of civil aircraft. The Soviets assisted in the preparation of the new guidelines. In fact, the USSR's Airman's Information Publication, issued to all Soviet fighter interceptor pilots, outlines the ICAO procedures to be followed and specifically details what signals to give an intruding civil aircraft at night: (1) Rocking wings from a position in front and normally to the left of the intercepted aircraft; (2) Flashing of navigational and, if available, landing lights at irregular intervals; (3) Making a general call on the emergency radio frequency 121.5 MHZ and repeating the call on the emergency frequency 243 MHZ.

Former Federal Aviation administrator J. Lynn Helms addressed an emergency ICAO meeting in Montreal, Canada, two weeks after the Soviets' destruction of KAL 007. "The evidence we have," he told the assembly, "indicates that the USSR fighters did not use these intercept signals before destroying KAL 007. It is clear from the communications that KAL 007 was unaware it was being intercepted."[4]

The radio intercepts made public in both Washington and Tokyo indicate clearly that the crew was given no warning. In fact, during the thirty-nine minutes of the recorded conversations of the Soviet fighter pilots, the word *target* is used over fifteen times to describe the jumbo jet.[5]

The Soviet fighter pilot who fired at KAL 007 claimed that he flashed his lights, wagged his wings, and then fired tracer rounds. The chief of the Soviet General Staff, Marshal Nikolai V. Ogarkov, claimed that the 747 "tried to escape" after the warning shots were fired.[6] A Soviet embassy official in Ottawa stated that the intercept transcripts belonged to some other air exercise not connected with KAL 007.[7] In Tokyo a Soviet embassy official openly denounced the transcripts as a fabrication.[8]

Immediately following the downing, there were six days of silence from the Soviet Union. Then they contended that the KAL airliner was a spy plane. An expert in Soviet disinformation had been at work on a cover story; he was none other than Marshal Ogarkov. He held an impressive press conference on September 9 that a *New York Times* column head called "Spellbinding Performance in Moscow."[9] *Newsweek* magazine reported that he was regarded by Western Soviet experts as the Moscow military's "main brain" and, as the Soviet Defense Ministry representative to the SALT negotiations, "he earned the respect of his U.S. counterparts and was viewed as one of the new, less rigid Soviet military leaders who gave Western observers hope for arms-control progress."[10]

A former high-ranking Soviet officer, who defected to the West and uses the cover name of Viktor Suvorov, wrote a revealing book in 1982 entitled *Inside the Soviet Army.* Unlike the writers for *Newsweek* and other Western journalists who attended Ogarkov's packed Moscow news conference, Suvorov was in a position to know what Ogarkov did before he became a close military associate of Yuri Andropov. He ran the Chief Directorate of Strategic Deception, the disinformation program for the entire Soviet military.

Suvorov writes:

> This Directorate makes a careful study of everything that is known in the West about the Soviet Union and fabricates an enormous amount of material in order to distort the true picture. . . . In addition, of course, representatives of the Chief Directorate, helped by Soviet military intelligence, have recruited a collection of mercenary hack journalists abroad, through which it spreads false information, disguised as serious studies. Its representatives attend negotiations concerned with détente, peace, disarmament, etc.[11]

Marshal Ogarkov and Moscow's worldwide network of operatives were moderately successful in making the case that the Korean airliner was a spy plane, especially after the United States admitted that an RC-135 reconnaissance aircraft had come within seventy-five miles of KAL 007. Moscow maintained that the 747 passed electronic intelligence data to the RC-135, itself equipped with sophisticated electronic intercept technology to perform intelligence gathering without the help of a civilian airliner.

Admiral Robert R. Inman, former director of the National Security Agency and deputy director of the CIA, pointed out that the Soviets' destruction of the 747 is a grim reminder about the Soviet value system that has little regard for innocent lives while conducting a sophisticated campaign to convince the world they are the peacemakers of the globe. Admiral Inman observed of the charges that KAL 007 was a spy plane:

> We have much better ways of getting the information. If the Koreans had had anything they had wanted to know, they would have come to ask us for it—not fly a 747 loaded with passengers when they're trying to build a commercial airline business. It's such a shallow lie that it makes the Soviet case even worse.[12]

One great failure of the Western news media is their lack of specific knowledge of the subjects on which they report. This was vividly illustrated during the coverage of the KAL 007 tragedy.

Aviation is a highly technical and specialized field. For example, the radio intercepts take on greater significance if read by a former interceptor fighter pilot. The average person reading the transcripts understood the words "I have executed the launch" (the missile) and "the target is destroyed." (In actuality, KAL 007 was not destroyed in the air but remained in one piece as it made a twelve-to-fifteen minute spiraling death plunge toward the sea.)

Not everyone reading the transcripts would understand that while the Soviets claimed the 747 was without navigation lights, the transcripts show that the Soviet SU-15 interceptor, designated as "805," stated on three separate occasions that the "target's A.N.O. [air navigation lights] are burning. The [strobe] light is flashing." Such rapid and intense bright lights are required for civilian aircraft because they can be seen by other aircraft at a great distance.

Colonel Samuel Dickens (USAF, Ret.) spent over twenty years almost exclusively in reconnaissance and fighter interceptor work in Europe, the United States, and Asia. After reading the radio intercepts recorded during the thirty-five minutes prior to KAL 007's being struck by two air-to-air missiles (aircraft to aircraft), Colonel Dickens was able to conclude that no doubt exists in his mind that when "805" lifted off from Sakhalin Island his mission was to destroy the 747. Colonel Dickens draws this conclusion based on several key elements contained in the radio intercepts.

First, at the very beginning of the transcripts we read "805"'s telling ground control that he has armed his missiles, a procedure clearly indicating intent to attack a target rather than warning it.

Second, a fighter prepared to fire a missile at an aircraft must first acquire radar and visual contact, which "805" did, and this contact was continued throughout the thirty-five minutes before launching his missiles from an attack posture by positioning himself behind and below the 747. None of the intercepts indicate that "805" was ever to the 747's left and parallel to the aircraft flight deck where he could be seen by the KAL pilot.

Third, the fighter pilot claimed later, as did the Soviet leadership, that he fired warning shots with his cannons containing tracer ammunition out in front of the 747 in the early morning darkness at thirty-two thousand feet. The flashes would have been seen, if tracers had been shot in front of the 747. Colonel Dickens observed:

> If he had a cannon in that aircraft, as soon as you pull the trigger you get very bright flashes and the pilot and copilot could not have missed those flashes. Peripheral vision is very important for pilots and you are in the dark, with the cockpit interior lights dimmed for night flying, and you have bright flashes in staccato fashion to your left, your head turns immediately. If he had tracers, as the Soviets claim, peripheral vision would not be needed because the crew would have seen the shells streaking ahead of the aircraft.[13]

The U.S. Air Force interceptor veteran also points out that "805" stated, "Now I'll try the rocket," indicating that he probably tried to down the airliner with his cannons first. The transcripts also reveal that "805" disarmed his missiles at one point and probably went in for a closer look from the rear and below, leading Colonel Dickens to conclude that ground control instructed him to go in closer and have a *second* look to make certain it was not a Soviet aircraft they were preparing to destroy. The clear night, a quarter moon, and the strobe light's flashing make it very unlikely that "805" did not know it was an unarmed civilian airliner he was stalking. In fact, U.S. intelligence sources in Washington later reported that on his debriefing on Sakhalin, "805" told his superior that he knew he had attacked a civilian airliner![14]

In the author's interview Colonel Dickens stated:

> Remember that a strobe light is lighting the entire aircraft, which an interceptor can see even if he is to the rear and below, and you know you have an enormous aircraft, even if in silhouette. When you come up on an aircraft of a 747's size, and I

> have as a fighter pilot, you are just awed by its size. Every air force in the world trains its pilots with drills to be able to instantly identify aircraft silhouettes. It is significant that the Soviets in Moscow claimed that their fighters did not get up high enough to see the hump of the 747 in silhouette, clearly indicating they never did get parallel and to the left of the Korean jumbo jet—otherwise there is no way they could have missed the hump.[15]

The radio intercepts reveal no hesitancy on the part of the Soviet pilot nor by his ground control as revealed by the pilot's responses to the control station. (Transmissions from the ground controls were not heard on the tapes.) KAL 007 was tracked for two and a half hours, first by as many as eight Soviet fighters out of Kamchatka and on radar, apparently to let the 747 proceed while instructions were requested from higher authority. The second group of eight fighters from Sakhalin Island apparently was given specific orders to shoot down the 747, which were executed just before the plane left Soviet airspace.

Colonel Dickens observed:

> In the twelve minutes of life after the missiles were fired, there probably was first a blinding flash, followed by decompression; and it probably took place in a matter of seconds. Explosive decompression would cut down the useful consciousness of the passengers, filled with panic in a darkened cabin brought suddenly awake and bewildered and unable to see or get the oxygen masks that automatically drop down. The crew would have had time to put on their oxygen masks and the pilot may have put the 747 into a steep dive. But the conclusion I come to is that the missiles went into the engines and the aircraft was not completely broken up.[16]

The missiles are called "heat seeking," and it is believed they entered the right jet engines, judged by the direction of the spiral viewed by the Japanese on their radar screen.

Japanese Self-Defense Force radar reported that the 747, after being hit by two missiles (confirmed by the radio intercepts of the fighter pilot), took twelve minutes to fall from thirty-five thousand feet to sixteen thousand feet (the aircraft had just reached thirty-five thousand feet after requesting from Tokyo Control permission to ascend from thirty-two thousand feet), before it dropped off the radar screen. However, a lower altitude radar station in northern Hokkaido established that the 747 had fallen in a series of wide spiraling loops, indicating that it was out of control at that lower altitude.[17]

Presumably the missiles, detonating below the right wing of the aircraft, rocked the cabin, probably twisting the fuselage, rupturing fuel tanks, and severing hydraulic lines so that the pilot had almost no control over the aircraft. At thirty-five thousand feet no explosive fire took place because the outside temperature was minus 58 degrees Fahrenheit, and the oxygen pressure was too low to permit combustion. Falling out of control, Captain Chun or Copilot Son shouted into their radio that they were experiencing rapid decompression. Tokyo Control asked the crippled 747 to try another frequency since its transmission was weak, almost unintelligible.

As KAL 007 spiraled toward its tomb in the frigid waters below, most of the passengers were undoubtedly unconscious. At approximately sixteen thousand feet the warmer air probably ignited a fire on the right wing where the jet fuel was gushing out, eventually causing an explosion, the bright blue flash that Japanese fishermen saw and heard southwest of Sakhalin Island.[18] In a matter of minutes the wreckage had reached the surface of the Sea of Japan and quickly sank out of sight to a depth of as much as two thousand feet with most of the 269 passengers and crew trapped and drowning, if they had managed to regain consciousness and survived the explosion and the impact.

Nine days later the first body to wash up on the northern coast of Hokkaido was that of an Asian child between six and eleven years old, its legs, right arm, and part of the top portion of its head missing. Japanese pathologists found small steel splinters in its head and glass and wood splinters in its chest, confirming the tremendous explosion heard and seen by the Japanese fishermen. Found also was the tip of the KAL 007 tail with the white letter "L" and part of what could be the letter "H." A Korean Airlines official, Park Chung-Hong, identified the fragment as coming from the tail section of a KAL plane.[19]

During his four years as a flight surgeon in the U.S. Navy, Congressman Lawrence McDonald had become an expert in air-sea rescue. He was perhaps the only one aboard KAL 007 with such specialized training who might have had a chance of surviving if there had been immediate rescue attempts. He was also the only one aboard the doomed airliner who understood completely the purposes and motives of Soviet-sponsored terrorism.

In August 1981 McDonald wrote:

> Terrorism may be properly defined as a violent attack on the noncombatant segment of the community for the purpose of intimidation, to achieve a political or a military objective. . . . The attention of the public is directed toward the acts of certain terrorists, but the controlling force and ultimate aims are more sinister and part of a global mosaic.[20]

13

The Impotence of Moral Outrage

"Not only did Soviet officials shoot down a stray commercial airliner and lie about it, they have callously refused offers of international participation in search and rescue efforts."[1]

Ambassador Jeane Kirkpatrick
U.N. Security Council
September 6, 1983

Sakhalin Island has always been a barren and isolated spot of suffering and death. Under the Russian czars it was a penal colony. In the latter part of the nineteenth century the Russian writer Anton Chekov visited the island and later wrote: "I have seen Ceylon, which is paradise, and Sakhalin, which is hell."[2]

The Russian novelist and Nobel Prize-winner Aleksandr Solzhenitsyn noted that, while "the name Sakhalin strikes terror," it was a far more brutal concentration camp—one of the thousands of Gulag slave labor camps—under Josef Stalin than when it was a penal colony under the czars.[3]

Swept and battered by ice storms in the winter and drenched with damp, wet weather in the summer, the narrow six-hundred-mile island was ceded to Stalin in 1945 by the allies after the defeat of the Japanese in World War II. It was not until well after Stalin's death in 1953 that Sakhalin was transformed by the Kremlin from an island of terror to a formidable military and naval base.

Closed to all but a few foreigners, there are still an estimated three thousand former Korean laborers residing on the island who were brought there from the then-Japanese colony of Korea. They have no official status even though most of their descendents have accepted Soviet or North Korean citizenship; and they yearn to return to their homeland, feeling they have been abandoned. And the Soviets refuse to allow them to emigrate.[4]

Solzhenitsyn maintains that the Soviet system can only be understood in terms of terror imposed from within and extended without; lies and violence are woven deeply into the fabric of the system that outlaws all regard for law. "For half a century and more," he wrote, "the enormous state has towered over us, girded with hoops of steel. The hoops are still there. There is no law."[5]

The shooting down of an unarmed airliner illustrated not only the Soviet Union's callous and brutal character, but also demonstrated the ease with which it was prepared to violate international law and agreements that it had signed with other nations as a member of an international organization. In the wake of the Soviets' shooting down Korean Airlines 902 in 1978, for example, Kremlin representatives helped draft new procedures in the International Civil Aviation Organization for the interception of intruding aircraft— only to ignore those same regulations when KAL 007 strayed, or was lured, into their airspace. As a member of the ICAO and signatory to several other agreements governing international civil aviation, the Soviets are obligated to provide search and rescue services and to permit other nations access to an airline crash site.

Federal Aviation Administrator J. Lynn Helms told a September 15, 1983, meeting of ICAO in Montreal:

> To date, the USSR has refused to permit search and rescue units from other countries to enter Soviet territorial waters to search for the remains of KAL 007. Moreover, the Soviet Union has blocked access to the likely crash site and has refused to cooperate with other interested parties to ensure prompt recovery of all technical equipment, wreckage, and other material that may facilitate and expedite completion of an investigation.[6]

In the two months following the destruction of KAL 007, the Soviets went to extraordinary expense, time, and effort to prevent outside recovery of any survivors, bodies, or wreckage just inside their own territorial waters, where it is calculated the plane sank. Immediately after the 747 disappeared from radar and fell into the Sea of Japan, the Japanese Maritime Agency received reports from its patrol boats near the general crash scene that five Soviet ships and two aircraft were criss-crossing the suspected crash site.[7]

Within forty-eight hours the Soviet ambassador to Japan, Vladimir Pavlov, told Japanese officials that search craft had picked up bits of wreckage but had no knowledge of any bodies' being found.[8] When Japanese and Korean search vessels came within hailing distance of Soviet ships, they were stopped by warning shots from entering Soviet territorial waters.[9] Repeated requests by Japan, Korea, and the United States to enter Soviet waters were ignored.[10]

In the first three weeks of the exhaustive search effort in international waters by the United States and Japan, parts of eleven different bodies of KAL 007 passengers were carried by the swift sea currents around Sakhalin to beaches on northern Hokkaido and found by fishermen and local police. As of September 24, fragments of the aircraft were also recovered, along with 187 shoes, 32 items of clothing, and 17 identification cards, mute evidence of the tremendous explosion that must have shattered the crippled airliner.[11]

The Soviets on September 26 turned over to U.S. and Japanese officials items that they claimed had washed ashore along the west coast of Sakhalin Island and the small adjacent Moneron Island, thirty miles to the southwest: seven pairs of trousers, a suitcase soaked in kerosene jet fuel, five battered oxygen bottles, six brown seat cushions, a torn life raft, and pieces of metal from an engine casing.[12]

The chief of the Soviet Border Forces for Sakhalin and Kurile Islands, Major General A. I. Romanenko, denied that his crews had recovered either the plane's cockpit voice and flight recorders or any human remains. Japanese foreign office official Minoru Tamba reported:

> The Soviets told us on the twentieth of this month [Sept.] they had no bodies or body parts to return. We asked Romanenko persistently, and I looked him straight in the face and asked if they really hadn't found any. Romanenko said the Soviets had found no human remains whatsoever.[13]

Eight times in the three weeks after the tragedy, Korean Airlines organized for the grief-stricken relatives a mournful memorial journey by ferry boat from northern Hokkaido to within a few miles of the crash scene so that the living could pay respects to their dead. The "mourning boat" of wives, husbands, sisters, brothers, and orphaned children let out anguished grief-stricken farewells amid the peal of Buddhist prayer bells and Christian burial services for the dead. Flowers, personal articles of clothing, and letters from children for their deceased parents were thrown into the swift sea. The *Soya Maru*'s deck rail, crowded with the bereaved, was the scene of tearful mourners calling out the names of their dead, one woman throwing a sweater into the cold waters off Moneron Island in the forlorn hope it would protect her dead son from the cold. "For twenty minutes," wrote a reporter later, "as the *Soya* drifted in the calm sea, with Moneron visible to the north, the promenade deck turned into an uncontrolled theater of tragic emotion."[14]

All requests to the Soviets to let the steady stream of mourning boats into the Soviet zone went unanswered. One woman who had lost her daughter tried to throw herself into the sea. Tearful relatives pleaded for the ship to get closer, only to be told by a Japanese Red Cross official: "No, we can't go any further. The Russians behave like a bad guy these days." Overhead, as the grieved relatives tried to hold a memorial service, a Soviet Ilyushin reconnaissance plane made no less than six close passes—the roar of its engines drowning out the mourner's sobs and calls to the dead.[15]

The Soviets mounted a massive sea and air operation in what was a clear effort to locate the KAL 007 voice and flight recorders and to prevent the world from knowing what actually transpired prior to the Soviet fighter's release of its missiles. At one time twenty Soviet ships were seen, some using deep diving gear and unmanned submarines lowered into the water. Located in the tail section of all civilian aircraft, the recorders emit a "pinging" signal for thirty days so they can be located in the event of a plane crash or accident.

U.S. Navy ships tracking the signal had experienced "electronic disturbances" in an apparent effort by the Soviets to disrupt the search; they later lost contact with the "pinging." A U.S. Navy statement also reported that Soviet ships "hampered U.S. salvage operations on several occasions by passing close aboard to and in front of our ships, thus presenting situations which could result in collisions and which required our vessels to take evasive measures."[16]

A U.S. naval officer described the search for the remains of the airliner, at depths of up to two thousand feet, "like trying to locate a pencil in the desert at night from an altitude of one thousand feet."[17] Despite initial optimism that the recorders might be found and reports that both the United States and the Soviets had retrieved them, within thirty days and the falling silent of the "pings" nothing was found officially.

Soviet efforts to impede the search efforts had become so extensive that the U.S. chief of naval operations filed a formal protest with Moscow, charging that the Soviets were not conforming to international law and the rules of the road at sea. Admiral James D. Watkins reminded the Soviets that in the 1970s an agreement had been reached between their two navies on the proper conduct of military ships when in close proximity during peacetime.[18] Moscow ignored the protest.

Twenty days after the formal United States complaint, a Soviet intelligence ship held a Japanese search vessel at bay for twenty minutes with its deck guns until an American ship arrived on the scene, warning the Soviet ship off by radio. The Pentagon sought to play down the incident, but one official was quoted as worrying that such Soviet behavior might presage a Soviet naval response if the United States found the wreckage and tried to bring it to the surface.[19]

The conduct of the United States in the weeks after the airliner was shot down was a sharp contrast to that of the Soviets. While the United States and its Western allies were making their case against Moscow in the international news media and the United Nations, the Soviets, the day after vetoing a U.N. Security Council resolution condemning their shooting down KAL 007, mounted a show of naval and air force might in the northern Sea of Japan on September 13. Soviet warships, using live ammunition, held war games in what was a clear effort to intimidate the Japanese. The Soviets even sent their aircraft within 186 miles of Tokyo to test Japanese radar defenses before they were turned back by Self-Defense Force jet fighters. "The Soviet attitude is brazen and unscrupulous," an angry Japanese Foreign Minister Sintaro Abe told reporters.[20]

United States Ambassador to the United Nations, Jeane Kirkpatrick, told a Security Council meeting prior to the Soviets' vetoing the resolution condemning the destruction of KAL 007, that it was a

> deliberate stroke designed to intimidate—a brutal, decisive act meant to instill fear and hesitation in all who observed its ruthless violence. . . . We are dealing here not with pilot error but with decisions and priorities characteristic of a system. Not only did Soviet officials shoot down a stray commercial airliner and lie about it, they have callously refused offers of international participation in search and rescue efforts in spite of clearly stated "International Standards and Recommended Practices" of the International Civil Aviation Organization, which call on states to "grant any necessary permission for the entry of such aircraft, vessels, personnel, or equipment into its territory and make necessary arrangements . . . with a view to expediting such entry."[21]

At a September 16 meeting in Montreal, the ICAO passed by a twenty-six-to-two vote a resolution "deploring," but not condemning, the Soviet shooting of KAL 007. The Soviets and Czechoslovakia voted against it, while China, Algeria, and India abstained. The resolution contained no sanctions.[22] South Korea's chief delegate to the ICAO Montreal meeting could hardly contain his smoldering rage at Soviet efforts to have the U.N. agency hold off taking a vote until the USSR could conclude its own investigation! "The resolution," Park Keun told reporters, "is far, far behind our expectations, to be frank. Nevertheless, as a member of the ICAO and the international community, we wanted to be responsible and accommodating."[23]

Unfortunately the term *accommodating* could summarize much of the activity of the West, when rage at what the Soviets had done was followed by impotent gestures that seemed only to embolden them. For example, in Tokyo, Canberra (Australia), London, and Washington, Soviet envoys refused to accept formal diplomatic notes demanding compensation from the Kremlin for the victims of KAL 007, each Soviet diplomat claiming that the United States was responsible for the deaths of those on the unarmed airliner![24]

At all of this, Congressman Lawrence McDonald would have been neither surprised nor angered, only concerned at how unhealthy it is for an individual, as well as for a nation and a civilization, to be consumed with outrage but hobbled by impotence born out of fear. The fundamental problem, he had said countless times during his career in Congress, was the inability of decent, intelligent people to believe they are personally, as a nation and as a civilization, threatened.

Shortly before his death, Congressman McDonald said:

> We must realize that we as United States citizens and as heirs to Western civilization should put our priorities in order and know that we are in a fight for our lives. . . . it's a matter of whether this entire civilization is prepared to fight the Communist plague or be destroyed. This is not something that our children or grandchildren will face. This is something that we shall have to face.[25]

14

Disbelieving Eyes and Ears

"A people are not capable and wise who feel extreme grief and anger for a while and then forget them with passing time."[1]

Chun Doo Hwan
President, South Korea
September 7, 1983

South Korea is a country hardened by a harsh history. Invaded and occupied by the Japanese for thirty-five years prior to 1945, the "Land of the Morning Calm" in June 1950 was also invaded by North Korea acting as a Soviet surrogate. The Korean War reduced the South to ashes. During the last thirty years the South Koreans have lived in a twilight state of neither war nor peace, while structuring a viable, prospering economy.

The last time the United States was invaded by a foreign power was by the British during the War of 1812—limited to areas along the East Coast. It has been almost 120 years since the end of the bloody War Between the States, even though its wounds never really healed until the earlier part of this century. The United States has no history of living with the constant threat of invasion by a rapacious power seeking submission and dominance by armed force.

The Koreans, Chinese, and Japanese know the Russians better than any peoples; bitter and bloody historical and contemporary experience has taught them to regard the Russian leadership for what it really is—an atheistic power elite with great feelings of inferiority that never developed a high, civilizing culture.

When the North American continent was inhabited solely by its native Indian tribes, when Europe was the campground of barbarian hordes, and when Russia was still a collection of feuding violent tribes, China, Japan, and Korea had developed separate but similar high cultures with value systems that put a premium on learning, loyalty, and codes of personal conduct, such as honor.

The Occidental West deludes itself when it accepts the Oriental East's passive outward exterior as a reflection of its interior. The mask worn by Orientals has been and remains a civilizing device, necessary in places where there are too many people and too little land. Not giving offense is the Oriental way of keeping in check fierce emotions and tenacious commitments that lie just beneath the surface.

United States fighting men in the Pacific during World War II found out firsthand just how fierce these emotions and commitments were when Japanese soldiers fought to the death rather than live in disgrace, dishonor, and defeat. Chinese and Korean soldiers during the Korean War demonstrated they could endure greater hardships and fight longer and with less material comforts than Western troops. The U.S. defeat in Vietnam was at the hands of fighting units inferior in firepower and logistics. The only military units during the Vietnam War that were consistently successful in defeating the North Vietnamese and Viet Cong were the South Koreans, who in a very short time pacified Quinhon province in Central Vietnam.

South Korea's ambassador to the United States on the night KAL 007 was downed was Lew Byong Hion. A veteran of the Korean War and commanding general of "the tiger" division in South Korean forces in South Vietnam, Lew was chairman of the Korean Joint Chiefs of Staff before his appointment as ambassador to Washington. While U.S. officials initially found it impossible to believe that the Soviets would deliberately destroy an unarmed airliner, Ambassador Lew on hearing the first reports that the 747 was missing, then reported to be safe on Sakhalin Island, said, "I knew, I had a fear and suspicion from the very start, that the Communists had shot the plane down."[2] Ambassador Lew personally knew KAL 007 pilot Captain Chun, remembering him as cool, quiet, and very professional. What most enraged Ambassador Lew about the Soviets was their dishonorable self-righteous attitude and callous contempt for the truth, alien to the values of a military man who is a descendent of fourteen generations of Korean teachers and scholars.

Ambassador Lew said with a slight smile:

> American people are so fortunate. You are so far away from Moscow, thousands of miles away. But look at us! We are situated against another Communist country, the distance from Seoul to the border only twenty-five miles. That is the equivalent of the distance between the White House and Dulles Airport. The Communists initiated the Korean War and turned our land into ashes, inflicting terrible casualties. In the last three and a half decades we have suffered. Yes! You people are fortunate.[3]

When Korean Airlines flight 015 landed in Seoul, none of its passengers noticed that KAL 007 was overdue. The uneasy feeling of Sen. Steve Symms on the flight from Anchorage and his brief but dark premonition of a Soviet attack were about to become horrible headlines of brutal and bloody facts.

"Immediately," Symms related to this author, "when I was told the plane was missing, I said, 'The Russians shot it down.'"[4] Fran, his wife, recoiled in disbelief, chiding her husband for thinking such dark thoughts.

Later while still in Seoul, Symms would recall encountering State Department officials and high ranking U.S. Army and Air Force officers who expressed the unanimous view that it was "impossible" to believe the Soviets would ever shoot down an unarmed 747 with so many innocent people aboard. A Marine Corps brigadier general, an aviator with numerous combat decorations, told Symms he could not believe that the Soviets would commit such a brutal act, suggesting in a friendly way that making such statements publicly might later prove embarrassing! (Symms is a former Marine officer.)

Frances Symms recalled:

> I didn't want him to be right, so I took the other side and played devil's advocate. When we heard that it was only forced down and everyone was safe, we felt so good about the news. I saw Steve later and he said to me and everyone at the hotel, including people from the U.S. Embassy, "Don't get your hopes up. I think the plane is in the water. I just don't feel good about it. There is just too much evidence pointing to the fact that something terrible has happened to that airplane. If they had been forced down on Sakhalin those rotten Russian SOB's would have been on the radio telling someone they had landed."[5]

As the day wore on in Seoul and the United States congressional delegation attending the conference came to realize that Symms was appallingly accurate, the warnings in the prepared speeches at the conference about the menace to the peace from North Korea and the Soviets took on a chilling life-and-death meaning. Senator Symms was asked by a reporter in Seoul what the United States should do. He replied with a question of his own: "What would the Israelis do?"[6]

Senator Orrin Hatch (R–Utah) correctly forecast that while Washington would call the Soviets liars, demand a U.N. Security Council session, and publicly condemn the Kremlin, no tough specific actions would emerge because "the State Department has already said no"[7] to specific kinds of actions previously advised by Sen. Henry Jackson and Congressman McDonald. He made that prediction only forty-eight hours after KAL 007 was shot down.

Senator Hatch would later tell this author that he believed the State Department was weak in its initial response, reflected in the fact that the Soviet ambassador in Washington was not summoned immediately to the State Department; instead, a lower Soviet diplomatic functionary was called.

> There was no real strong activity. I personally believe that if the Japanese foreign minister had not spoken out so forcefully in the beginning, we would not have gotten out of the State Department the response we did, which should have been stronger. Our State Department looks at matters through the

> eyes of the other world, rather than through the eyes of the United States. I think some of these people become paralyzed if they have to make a decision when it might have to be tough.[8]

Hatch was in Tokyo when he first learned that KAL 007 had been shot down. The Japanese politicians were no less late in catching up with the wave of angry public indignation than those in Washington. However the difference in Washington and in Tokyo was significant. In Washington little public attention was paid to the question of Soviet insincerity, whereas in Tokyo official and unofficial circles, talk was about nothing else.

New York Times Tokyo correspondent Clyde Haberman wrote:

> One of the things that nettled the Japanese was the Russians' lack of "sincerity" as they rejected blame for attacking the South Korean plane with heat-seeking missiles. "Sincerity" came up almost daily. It is an important constant in Japan, and even if it involves nothing more than a ballplayer offending his manager, a "sincere" act of contrition takes care of a multitude of sins.[9]

The key to the South Koreans' reaction was not only angry rage at the Soviets for their act of airborne terrorism; a silent and smoldering rage began to build among the Koreans when they realized that the United States was prepared to do so little so quickly. Like the Japanese, the Koreans understood the deeper motives behind Moscow's actions against the Korean airliner—that they went beyond mass murder and involved the Kremlin's use of terror to strike at the United States through a smaller and weaker ally. "Well, we might wish to see," acknowledged South Korea's U.S. ambassador Lew, "more, stronger, harsher action; but we know very well every foreign nation has its own domestic concerns and conditions."[10]

President Reagan's September 5 television speech ended by his announcing that the United States would continue arms reduction talks with the Soviets. "We know," Mr. Reagan said, "it will be hard to make a nation that rules its own people through force to cease using force against the rest of the world. But we must try."[11]

Two days later, South Korean President Chun Doo Hwan ended a statement, read at the Seoul memorial service for the victims of KAL 007, on a theme that may have been meant as much for the Americans as for the Koreans.

> We should pledge today to stand in the vanguard of efforts to realize the wishes of the dead and should reflect on how to genuinely console them. A nation that does not fight against violence and injustice is a nation without life. A people are not capable and wise who feel grief and anger for a while and then forget them with passing time. A man can dwell on sorrow and resentment for a long time, but a nation should not. A clever nation does not let a tragedy end in tragedy, but with renewed energy uses it as a goad. We must have keen insight to see what we are and where we stand to reexamine what it is we should do.[12]

What the South Koreans did and said in the aftermath of KAL 007 provides a clue to their character as a people, just as what the United States did and said offers the same insight. When the KAL 007 memorial service began in a packed Seoul stadium, attended by angry thousands, sirens wailed all across South Korea as a signal for one minute of prayer for the crew and passengers.[13] Thousands of Koreans in dozens of cities vented their angry rage in demonstrations, many of them doubtlessly organized by the government. But by contrast there is evidence, as will be shown, that U.S. political leaders, the news media, and many of the American ruling elite made conscious efforts to manipulate and defuse the anger and rage of the American people out of fear of the Soviet Union.

In the United States there was no plan for a memorial tower to the victims of KAL 007, whereas the Korean government discussed erecting such a memorial at Kimpo International Airport outside of Seoul.[14]

No U.S. clergyman composed a song as a prayer to the victims of KAL 007 as did American Roman Catholic priest Father Raymond Sullivan in Korea for what he explained as "one of my prayers to console the souls of the victims of the ill-fated jetliner."[15]

No U.S symphony orchestra premiered a special orchestral song composed in memory of the victims of KAL 007, as did the Seoul Philharmonic Orchestra under the baton of the German conductor Walter Gillessen on October 7. Composed by forty-five-year-old Professor Pak Chun-sang and sung by the noted South Korean soprano Teresa Song, "Requr Eskat In Pace" (Praying for Eternal Life) was as much a hymn of remembrance and benediction as it was an anthem of defiance. "I was indignant," Dr. Pak explained, "to hear of the Soviet barbaric attack and then inspired by my artistic instinct to pray for the victims." [16]

No American Catholic cardinal raised his voice in warning of the terrible future meaning of the airline massacre as did Korean Catholic prelate Stephen Cardinal Kim Sou-hwan. "I am afraid," he warned, "that a similar massacre or other tragedies beyond our imagination could happen throughout the world in the future if we fail to honor the preciousness of human beings."[17]

Twenty days after Cardinal Kim's remarks were published in Korea, a bomb blast went off in a Rangoon, Burma, cemetery, killing seventeen cabinet and top level officials of the South Korean government and four other persons. The blast narrowly missed by minutes killing the president of South Korea who was on a state visit. A little less than a month later the nominally neutral and pacifist Burmese government confirmed what the South Koreans had insisted on from the onset: the mass assassination effort was the work of North Korea. A November 4 Burmese government statement, after an exhaustive investigation and interrogation of two of the bomb plotters, "firmly established" that three North Korean military personnel—a major and two captains—had planted the bomb. Burma, as a consequence, broke diplomatic relations with North Korea and ordered all its diplomatic personnel to leave Rangoon in forty-eight hours.[18]

The U.S. news media virtually ignored a letter written by a nine-year-old Hong Kong girl to Soviet leader Yuri Andropov. Choi Man-yee wanted to know why her friend, eight-year-old

Yuen Wai-sum, and her parents had to perish on KAL 007. The girl said she was writing to Andropov because earlier in the spring of 1983 he had answered a letter from eleven-year-old Samantha Smith in the United States and even had invited her to the Soviet Union to prove he was a man of peace. "I am heartbroken," she wrote to Andropov who did not reply, "and have nightmares because my best friend is dead. . . . You are the leader of the Russians. Can you tell me why the Russians had to kill her?"[19]

Perhaps a larger question is why the United States so swiftly went back to business as usual with the Soviets.* Senator Symms may have provided part of the answer when he revealed how Sen. Jesse Helms had queried the State Department as to whether the U.S. congressional delegation in Seoul should return home by commercial airliner or by special Air Force jet as a security precaution.

> Our State Department Task Force reviewed Senator Helms's question and replied they thought it would be safe for us to return by commercial airliner since they believed that the Soviets had mistakenly shot down the airliner! I had a fit with our ambassador in Seoul, "Dixie" Walker, when he told me what they had said. The Soviets don't have to use disinformation, they just get their own propaganda line from our State Department. When we got back to the U.S. that is exactly the line the Soviets were putting out, it was all an accident.[20]

Five years before his death, Congressman Lawrence McDonald had come to the conclusion that the United States had ceased to be the leader of the free world, citing evidence of the two no-win wars—Korea and Vietnam—and the abandonment of a growing list of allies. As a consequence, he proposed a free world anti-Communist alliance—a fourth world of nations—made up of Israel, South Africa, Brazil, South

*It might be argued that it was a Korean plane, not a United States plane; that there were many more other nationals aboard than Americans. But the same lack of a strong response was clearly manifest when a terrorist's truck bomb destroyed the lives of over 230 American Marines, Naval and Army personnel about six weeks later in another part of the world.

Korea, and other nations placed in peril by the efforts at détente with Moscow and Peking.

McDonald's proposal had been made before Rhodesia (now Zimbabwe), Iran, and Nicaragua turned openly hostile to the West, without a concerted effort by the United states to prevent it from happening. In all three of these instances the United States shares a degree of responsibility for the destabilization and subsequent overthrow of governments once firmly our friends. These three countries are now anti-U.S. and pro-Soviet.

Congressman McDonald wrote in March 1978:

> Painful as it is for any American to admit, our country's so-called leadership of the Free World has been the leadership of the Judas goat taking the sheep to the slaughter. If it is not sufficiently apparent today, we can predict in full confidence that it will be more apparent tomorrow. Realizing the truth of this is half the battle, because it means the shattering of a whole frame of reference. Yet those leaders of countries not yet enslaved, if they are realists and brave men, can fight through disillusion and despair and see that they must make their own arrangements now. They can best do that by joining with others similarly situated, to multiply their strength rather than waiting their turn to go down with a whimper.[21]

15

The Dishonoring of the Dead

"It was a disgrace that so little was said and done on Larry McDonald's behalf."[1]

Sen. Orrin Hatch (R–Utah)
September 26, 1983

Eleven days after the Soviets destroyed KAL 007, the Kremlin's state-controlled news media expressed the conviction that worldwide revulsion would be brief and would not affect what one Moscow television commentator termed "the structure of détente."[2] That cynical conclusion was well founded.

On September 1 at 10:45 A.M., Secretary of State George Shultz appeared in the State Department's press-briefing auditorium before a crush of correspondents and a battery of television cameras. In a matter of a few minutes he gave what veteran Washington correspondents described as the most detailed diplomatic summary of an international crisis in U.S. history.

The Soviets had not only tracked KAL 007 for two and a half hours with their radar and at least eight interceptor fighters, Secretary Shultz told reporters, but the Soviet fighter pilot who shot the aircraft down had visual contact and knew it was a civilian 747 and had deliberately destroyed the "target" with a heat-seeking missile. "The United States reacts with revulsion to this attack," Shultz told reporters and the nation, barely able to control his rage. "Loss of life appears to be heavy. We can see no excuse for this appalling act."[3]

In follow-up questions Shultz was asked whether he had any theories about Moscow's political motivation for the massacre. "I can't imagine any political motivation for shooting down an unarmed airliner," he replied.[4] In the weeks to follow, the State Department, the White House, and a majority of members of Congress would consistently side-step the question of Soviet motivation. From the first hours after it was confirmed that KAL 007 had been blown out of the air, a carefully orchestrated effort was set into motion to manipulate and diffuse public anger and demands for tough retaliatory action against the Kremlin while at the same time continuing business as usual. It was a masterful exercise of saying much and doing little.

President Reagan was said to have been out of direct touch with the Korean airliner crisis for almost a full eight hours while vacationing at his ranch in Santa Barbara, California. Presidential Press Secretary Larry Speakes maintained that aides did not want to intrude on Mr. Reagan because of the difficulty of getting a clear picture of events, although they knew by 3:00 A.M. Eastern Daylight Saving Time (midnight in Santa Barbara) that the plane was down.[5]

Considerable criticism was leveled at the president for what appeared to be confusion, indecision, and a lack of leadership during the first forty-eight hours of the KAL 007 crisis. A more accurate estimate, based on what transpired subsequently, would be that the president, his White House advisors, and the State Department spent the first two days, which appeared to be dominated by indecision, ruthlessly planning a strategy to control and direct the crisis rather than allowing the crisis to control the administration to the point that it would be forced to take decisive measures against the Soviets. The administration's worst fear was that KAL 007 would escalate into a super-power confrontation, even armed conflict. It was fear combined with cold-blooded domestic political considerations, not moral outrage and a sincere desire to see justice done in the name of the dead, that animated the Reagan administration's response to the airliner tragedy. The evidence clearly leads to no other conclusion.

On September 2, for example, the day after President Reagan personally spoke to Mrs. Kathryn McDonald and expressed his outrage but not his support for strong retaliatory measures, he read a statement to reporters at Point Mugu, California. Mr. Reagan began by saying how "deeply saddened" he was to learn that Sen. Henry Jackson (D–Wash.) had died suddenly on the same day KAL 007 had been shot down.* "He was a friend, a colleague, and a true patriot, and a devoted servant of the people," the president said, adding, "He will be sorely missed and we both extend our deepest sympathy to his family."[6]

No mention was made of Congressman McDonald! Mr. Reagan, in subsequent public statements would provide the strange spectacle of expressing outrage and anger at the Soviets over their downing of the airliner while finding it impossible to mention the name of the man who had for his entire political career warned that Soviet Russia was an evil empire.

The following day, September 3, Kathy McDonald sent a telegram to Mr. Reagan inviting him to speak at a memorial service at Constitution Hall in Washington organized by the Conservative Caucus, a grass-roots organization that the Georgia Democrat had supported and served as a board member. Caucus Chairman Howard Phillips noted:

> Frankly, I was rather shocked that the president failed to mention Representative McDonald in any of his statements. The reason we invited President Reagan is that Representative McDonald is, as far as we know, the first member of Congress ever shot down by the Soviet Union. We thought it would be appropriate for the president to comment.[7]

No comment, however, was forthcoming and not for the lack of opportunity. In his September 5 nationwide television address the president did say at the very beginning that

> our prayers tonight are with the victims and their families in their time of terrible grief. Our hearts go out to them—to the brave people like Kathryn McDonald, the wife of the con-

---

*Senator Jackson suffered a fatal heart attack immediately following a press conference in which he bitterly denounced the Soviets for the KAL 007 massacre.

gressman, whose composure and eloquence on the day of her husband's death moved us all. He will be sorely missed by all of us here in government.[8]

This effort to placate McDonald supporters and the conservatives only made matters worse when later on during his TV talk the president delivered an effusive eulogy about Senator Jackson (a more moderate anti-Communist than Congressman McDonald). Particularly infuriating to his conservative supporters was his announced intention to proceed with arms reduction talks with the Soviets while imposing mild, if not meaningless, sanctions on the Soviets, centered principally in the area of aviation and cultural agreements.

New York Times columnist William Safire, a Reagan supporter and former speech writer to President Richard M. Nixon, said that Mr. Reagan's strategy was essentially propagandistic in an effort to undercut his critics and collect votes in the House of Representatives for passage of his defense proposals. Safire wrote:

> Never in the course of presidential history have so many bombastic words been accompanied by so much handwringing and such little action. No wonder dovish commentators have been lavishing praise on Mr. Reagan's decision to limit his reaction to an orgy of oratorical self-rightousness: no matter what the provocation, the march to the election-year summit must go on.[9]

Mr. Reagan, in fact, told *Time* magazine in an interview after his TV speech that despite KAL 007 and Soviet lying he would not rule out a Reagan-Andropov summit! "If I can be convinced in my mind," the president said, "that a summit can be beneficial to our security, to the free world, then such a summit should take place."[10]

William H. Gregory of *Aviation Week & Space Technology,* one of the few American publications that produced a detailed and accurate report of the airline tragedy, saw Mr. Reagan's mild response as a contrast to his warmongering image painted by his political enemies.

> The picture of the U.S. sitting down at the arms control negotiating table to negotiate in good faith with the Soviets is revolting now. The administration is paying more attention to the 1984 campaign politics than to the question of whether an agreement with the Soviets is more than a scrap of paper.[11]

The dictates of domestic politics clearly made Congressman McDonald a forgotten and shunned figure, not only by the White House but by a majority of his colleagues in Congress. There were a few exceptions. For instance, Georgia Republican Congressman Newt Gingrich made the ironic observation that McDonald, a Democrat, raised money for Republican candidates in the 1980 election that was critical in electing conservatives and defeating liberals, which gave the Republicans control of the Senate for the first time in decades. "I think," he told this author, "had he not been out there raising money in 1980, we would not have won control of the Senate for the Republicans." The former history professor added:

> Larry was a loner. He was a man driven by the intensity of his vision of the crisis of the West. He was totally committed to freedom; he had a very long-range sense of history, and he was willing to do the fundamental things that were starting to pay off. I think had he survived that he would have been seen in history as someone who, in fact, in some fundamental ways, changed the debate in this country. Frankly, I think he was ineffective in the House, but I think he was far more effective in the country. He understood the cost of doing what he thought was right in Washington, and as a patriot he thought it was his duty to do so.[12]

The death of Sen. Henry Jackson gave both the Reagan administration and members of Congress an excuse to ignore McDonald. Sen. Orrin Hatch (R–Utah), generally approving of Mr. Reagan's "measured response" to the KAL 007 massacre while contending it could have been stronger, was unsparing in his criticism of those in Congress and the administration who found themselves incapable of memorializing McDonald in the way they did Senator Jackson.

> Larry was a decent and honorable, honest, hard-nosed man. He stood up to his own liberal party and had primary oppo-

> nents from within his own party dominated by liberals because of his fight against communism. He was a courageous and honest man who deserved a more lasting memorial than a few mentions in the *Congressional Record.*[13]

For months before KAL 007 the Reagan White House's political game plan had been involved in making him appear a man of peace. As cold-blooded, and even cruel, as it appears, it was vital at all costs for the president to distance himself from McDonald and his supporters. It was for this reason that Mr. Reagan's advisors successfully urged him not to attend the McDonald memorial service in Washington. It would have reinforced the image of him drawn by critics as a right-wing warmonger. "The president is coming out of this looking good," said one State Department official questioned by *Newsweek* magazine. "He isn't looking like a crazy warmonger, and Shultz deserves a lot of credit."[14]

When a firestorm of criticism erupted from the conservatives about what they perceived as a callous attitude toward McDonald in particular and the dead in general and the weak response of Reagan pushed by the State Department, the White House, according to *Newsweek,* "appeared to relish the situation—believing conservative criticism only makes Reagan look more moderate and responsible as the 1984 race heats up."[15] Howard Phillips of the Conservative Caucus used the analogy that Mr. Reagan "sounds like Winston Churchill and acts like Neville Chamberlain,"[16] indicating that the president was guilty of more than rhetorical overkill.

Apparently stung by the criticism of his most staunch conservative supporters, who are united by the creed of anticommunism—of which McDonald was its most eloquent and intelligent advocate—Mr. Reagan uncharacteristically lost his cool. "I know that some of our critics," he said on September 9, "have sounded off that somehow we haven't exacted enough vengeance. Well, vengeance isn't the name of the game. Short of going to war, what would they have us do?"[17]

Two days later to a packed Constitution Hall, during the memorial service for Congressman McDonald, Howard Phillips answered Mr. Reagan with a telling on-target reply: "Mr.

President, it is not vengeance which we seek, but simple justice and godly retribution."[18]

The Washington establishment of Democrats and Republicans dishonored the 269 victims of KAL 007 because they ceased to view them as victims. Instead, they regarded them as a body of conclusive evidence that contradicted their most deeply held convictions that, in the final analysis, the Soviets were *not really* an evil, brutal empire.

Georgia Republican Newt Gingrich pointed out that the difference between Congressman Lawrence McDonald and Sen. Henry Jackson, besides the fact that one died of a heart attack and the other was murdered by the Soviets, was that McDonald was forever rudely reminding the Washington establishment of the costs of their behavior.

> The liberal establishment is as unhappy with the evidence of Soviet brutality as the liberal establishment in England was unhappy with the evidence that Adolf Hitler was a thug. And, therefore, the second they can forget it, they will. The net result is that you are very unlikely to have a minitelevision drama about the Korean airliner. More effort will be spent on two nuns being killed in El Salvador than on 269 people being killed by the Soviet Union because, if they discovered the Soviet Union was that bad, they would be forced to change. There is a psychological process that says people will go to any lengths to avoid having to change their opinions.[19]

16

Two Cultures in Collision

"If you don't stand for something, you will fall for anything."[1]

Rep. Lawrence P. McDonald
Recalled at Memorial Service
Washington, D.C.
September 11, 1983

How a nation honors its living and its dead is a snapshot of its character, values, and state of mind.

Two memorial services were held, within forty-eight hours, in Washington, D.C., for the victims of Korean Airlines Flight 007. Each of these services said significant things about America in the 1980s.

The first memorial service was held at Washington's National Cathedral on September 9. Attended by President and Mrs. Reagan, members of his cabinet, and seventy-five members of families that lost loved ones on KAL 007, the service became a sound stage for the White House to communicate to an angry and confused America that the administration was doing the decent and Christian thing by holding the service and declaring September 11, 1983, a National Day of Mourning. By doing so, it relieved President Reagan of the obligation of attending a memorial service on the National Day of Mourning at Washington's Constitution Hall organized by the conservative supporters of Congressman Lawrence McDonald.

President Reagan told the families of victims:

> No words can compensate for the burden of sorrows you carry. At times like these we can only trust in God for His mercy and wisdom. We are determined to do everything we can to see if

> things can't be done so that events like this will never happen again. I promise we'll do everything we can.[2]

Bishop John T. Walker played his part in the religious tableau perfectly when he praised President Reagan for his "controlled anger," offering the view in his sermon that the context for the brutal Soviet behavior was the result of distrust and fear. "When the tears are over," he told the 750 mourners, "when the anger has subsided, we must continue the negotiations out of which we further pray that peace will emerge."[3] The singing of the hymn "A Mighty Fortress Is Our God," composed at a time when Christianity was more courageous, seemed out of place with Bishop Walker's words that amounted to the theology of appeasement.

The men and women who joined President and Mrs. Reagan at the National Cathedral represented, with perhaps the exception of the grieving families, members of the Republican establishment who had demonstrated a capacity for remaining out of touch with the realities of the world. They were the product of a lifetime of wealth that produced blurred, out-of-focus beliefs, making it easy for them to adopt liberal, conventional wisdom despite Mr. Reagan's past hard-line rhetoric aimed at the Soviets.

Columnist George Will, a Reagan supporter, suggested that the public mind is like wax—easily shaped when it is heated—and that the president missed an opportunity to shape it, even laboring to minimize the opportunity.

> The Korean airliner atrocity raised the public's temperature to a healthy degree. But Reagan has squandered the moment, using it to solve what he evidently thinks is one of his political problems—a problem that he is not as peace loving as the editors of the *New York Times*. In the process he has dissipated a national asset: the Kremlin's anxiety that he just might mean what he says.[4]

Every action of the administration during the first three weeks of the KAL 007 crisis clearly communicated to the Kremlin that it had easily gotten away with mass murder. It is significant that a few days after the National Cathedral memorial service the state-controlled Soviet media was openly saying that the KAL 007 incident would not affect "the structure of détente."[5] The preoccupation with preserving the image of Mr. Reagan as a peace president was clearly indicated when *Newsweek* quoted a senior White House aide as saying, "A year from now people will say that the hipshooter kept his gun in his holster when he came under fire."[6]

It was out of such concerns that Mr. Reagan's advisors were successful in persuading him not to attend the Constitution Hall memorial service. The weekly conservative publication *Human Events,* said to be regular reading for the president and thus on good terms with the White House, reported that Mr. Reagan stayed away because: "The White House apparently feared a verbal assault from the podium that the president would be unable to answer."[7]

The fact that the White House did not even send an official representative suggested strongly that more than fear of a confrontation with conservative critics was behind the decision. Rather, White House aides concluded in cold political terms that it would have destroyed Reagan's carefully cultivated image as "peace president" to make an appearance at Constitution Hall before conservative supporters who were demanding tough action against the Soviets.

Almost four thousand grieving friends and supporters of Congressman McDonald packed Washington's Constitution Hall on September 11. Many were lower and upper middle-class conservative activists, most of them the product of working class families.

Three Reagan administration officials did come as private citizens out of concern and conviction: Morton Blackwell, special assistant to the president for public liaison; Donald Devine, director of the Office of Personnel Management; and J. William Middendorf, ambassador to the Organization of American States. A handful of members of Congress put in an appearance, while the bulk of the mourners were representatives of the fundamentalist Christian political right, who, like McDonald, scorn politics without principles and persist in the increasingly unpopular belief that the Kremlin leadership is a satanic force that cannot be fought without religious faith. "Their grief," wrote journalist John Rees later, "was transformed not into despair but into a terrible anger and determination that the Soviet Union must be held accountable for this crime against humanity."[8]

For almost three hours Constitution Hall rang with denunciations of the criminal conduct of the Soviets and demands that the Reagan administration take measures that would hold the Soviets accountable for their crime. Howard Phillips, chairman of the Conservative Caucus and organizer of the memorial service, proposed nineteen measures that the administration game plan did not include—ranging from ending aid and trade to breaking diplomatic relations.[9] Fear, according to Phillips, was the foundation of the administration's inaction—fear of appearing immoderate, fear of losing profits in trade with the Soviets, fear of public opinion, fear of confrontation, and fear of conflict itself.

Six days later a *New York Times*–CBS News poll provided support for many of the conservative positions. Of Phillips's nineteen action items, for example, the first was the release by the administration of all the radio intercepts between Soviet pilots and USSR personnel. The poll reported that 61 percent believed the administration had not told the entire story! Also 63 percent of the 705 adults in the poll favored canceling the grain sales to the Soviets. But the poll also revealed that a majority did not favor breaking diplomatic relations or breaking off the arms control talks.[10]

A *Newsweek* magazine poll taken by the Gallup organization revealed that a majority believed Mr. Reagan had not been tough enough with the Soviets while favoring, by 74 percent and 64 percent respectively, restricting loans and credits to the Soviets and reducing tourism between the two countries. But the poll also found 70 percent against cutting off the arms talks.[11]

The president's own pollster, Richard Wirthlin, also found that people believed that Mr. Reagan should have taken a tougher stance initially. White House deputy press secretary Larry Speakes revealed that of the phone calls to the White House in the immediate aftermath of President Reagan's September 5 television speech, 1,526 callers were angry that Mr. Reagan had not been tougher, while 734 calls supported the president.[12]

The evidence, therefore, indicates that the Reagan administration misjudged the gut reaction of a broad cross section of Americans—just as it misjudged the angry demands of the conservatives at Constitution Hall as representing only extreme right-wing views of a minority. How does one explain the persistent pursuit of the Reagan administration's policy of accommodating the Kremlin in the face of popular demands for stronger measures against the Soviets?

Congressman Newt Gingrich (R–Ga.) maintained that public anger and the passive policy of our leaders over KAL 007 underscore the largely ignored reality that U.S. politics is not determined by popular will but by the minority power of a liberal elite.[13]

It was against this elite that Congressman McDonald waged a war of words. In return he was written off as an extremist; and his membership in the John Birch Society settled the conviction of most that he belonged to the lunatic political fringe—without their troubling to read what he had written, to listen to what he was saying, or to analyze the reasons he gave for assuming the position he did during his political career.

During the Constitution Hall memorial service the stereotyping of McDonald (one of the few congressmen who opposed a national holiday for Martin Luther King, Jr.) was shattered when the Reverend Imogene Stewart, a well-known black evangelist from Georgia, was introduced.

John Rees wrote:

> She provided a gentle and appropriate personal remembrance, telling of her appreciation of the standards which Larry McDonald raised. She recalled, how, when "liberals" would not receive her, Congressman McDonald took the time to explain how Congress operates and to assist her in many ways. She remembered his saying to her: "There is one thing you must remember: If you don't stand for something, you will fall for anything." With that she sang *a capella* the beautiful hymn

> "Amazing Grace," and thousands of Larry McDonald's friends wept unashamedly.[14]

The mournful high-pitched strains of a lone bagpiper, from the clan Donald, filled Constitution Hall with the strains of "Going Home" and "Taps." Everyone with a sense of history was reminded that it was McDonald's ancestors, the Scots, who, under the religious rebel John Knox, had launched a long series of events that eventually led to the establishment of the United States by the Puritans and Calvinists.

Historian Otto J. Scott observed to this author:

> Larry McDonald was a convert to Calvinism. The American Calvinists came down from John Knox, who believed it was the duty of the Christian not to endure an anti-Christian or un-Christian regime. Knox believed it was the duty of every Christian to fight. When Larry converted he became very aware of the significance of his theological position.
>
> The Soviet Union has managed in its empire to break every dissident except those who have a strong religious faith. The KGB has admitted that there is no point in torturing the true believers because they cannot be broken. And, therefore, the Kremlin is well aware that its most serious challenge comes from religious groups.[15]

17

The Shipwreck of a Strategy

> *"I think it would be a marvelous idea if the United Nations were moved to Moscow."*[1]
>
> Burton Pines
> Vice President
> The Heritage Foundation
> September 20,1983

When Lawrence McDonald and 268 others perished on KAL 007, the Reagan administration decided that it would make its case against the Soviets at the United Nations. The results of that effort were not only meager, but momentarily there was a groundswell of public support that the international organization move out of the United States and that the United States get out of the United Nations.

On September 1 the United States, in association with the Republic of South Korea, sent a letter to the president of the U.N. Security Council, requesting an urgent meeting to consider the KAL 007 massacre.[2] The next day, in the first session, the Soviets insisted the meeting was unnecessary and that the United States had called the meeting as "only a cover, a counterfeit coin tossed down in the dirty game of anti-Soviet policy they are playing."[3] Quoting the Soviet news agency Tass, Richard S. Ovinnikov gave the first hint of the emerging spy-plane scenario that the Kremlin would follow by pointing out that the "relevant U.S. services followed the flight throughout its duration in a most attentive way."[4]

Charles M. Lichenstein, deputy U.S. ambassador, would later reply, "No, I would remind the representative of the Soviet Union: we followed you following the flight."[5] The Soviets even went further, insisting that somehow the United States desired the destruction of the 747 so as to reap propaganda benefits. Ambassador Lichenstein replied, following the Soviet argument to the logical question: ". . . were it to redound to the benefit of [our] administration, why, I must ask the Soviet representative, did you shoot down the airplane?"[6]

Canada's U.N. ambassador, joining with the United States and others in bluntly condemning the Soviets' action as wanton, calculated, deliberate murder, pointed out that "any effective action" the Security Council must undertake should not be hindered by a Soviet veto. "Any tactic of that kind," added Ambassador Gerard Pelletier, "would be considered as a lack of conscience and would quite rightly be interpreted as an admission of guilt."[7] After several lengthy sessions of the Security Council, the Soviets did in fact veto a compromise resolution on September 12; it did not abstain from voting. It was the 115th veto for the USSR compared to 36 by the United States, since both powers were given the veto exclusively when the U.N. was first established in 1945.[8]

Without a doubt the Reagan administration mounted an impressive, documented indictment against the Soviets, led by U.N. Ambassador Jeane Kirkpatrick and her deputy. Very early, however, the strategy of going on the offensive started to come unraveled. On September 7, for example, the United States was warned by Third World U.N. Security Council members that the desire to denounce the Soviets by name in any draft resolution might cost the United States votes. Put another way: they would vote for a resolution that denounced the crime so long as the criminal was not specifically named.[9]

Nine nations—a bare minimum needed for passage—voted for the resolution, which expressed how "gravely disturbed" a majority of Security Council members were that the 747 had been shot down, and the resolution "deeply deplores the destruction of the Korean airliner and the tragic loss of civilian life therein."[10] There was no outright condemnation of the Soviets, although the resolution did mention that KAL 007 had been shot down by a Soviet military aircraft.

New York Times U.N. correspondent Bernard D. Nossiter wrote:

> The result of the voting in the United Nations Security Council spared the United States what had threatened to become a political humiliation. As late as this morning [September 12] there were only eight certain votes for the resolution. Had the total gone unchanged, the document would have failed; and the negative Soviet vote would not have counted as a veto.[11]

Socialist Malta was persuaded to join as the ninth and critical vote, after expressing reservations that circumstances surrounding KAL 007's destruction "have been clouded with too much uncertainty."[12] While the Soviets and Poland voted against the watered-down resolution, China, Nicaragua, Zimbabwe, and Guyana abstained.

In 1980 presidential candidate Ronald Reagan had been critical of the Carter administration's appeasement of terrorists and of Third World countries at the United Nations. But once in power, while the rhetoric remained hard, the policy remained soft. When Ambassador Kirkpatrick and a few others within the administration urged that some $75 million in U.S. aid to Zimbabwe be slashed by 50 percent for its U.N. abstention, Secretary of State George Shultz vetoed the proposal, insisting that Zimbabwe was a key to U.S. policy when it came to pressuring South Africa to surrender control of southwest Africa.[13]

As with the Carter administration, the Reagan White House and State Department were prepared to let policy be fashioned in part by domestic political considerations. Congressman Julian C. Dixon (D–Calif.), chairman of the Congressional Black Caucus, for instance, argued in a letter to Secretary Shultz that continued aid to Mugabe was "fundamentally important to our relations with Africa, and it makes no sense to compound the tragedy of the Korean airliner incident by hurting relationships with Zimbabwe."[14]

President Reagan sought to maintain that the Security Council's watered-down resolution represented a substantive achievement of the administration over the KAL 007 massacre. A State Department spokesman was reported as saying that the United States was pleased with the U.N. vote.

"Although the Soviets did veto it, the resolution did express the world community's revulsion at the plane's destruction and its rejection of the Soviet explanation," Alan Romberg of the State Department said.[15]

The Soviets, however, had a far better understanding of how to manipulate propaganda levers at the United Nations than did the United States. While President Reagan and the State Department were maintaining that the Soviets were isolated at the international organization, the Kremlin counterattacked by announcing that Soviet Foreign Minister Andrei Gromyko would boycott the opening session of the U.N. General Assembly on the pretext that the United States had not guaranteed his safety or the proper handling of his airplane.[16] Ambassador Kirkpatrick pointed out that the Soviets' decision was motivated "for propaganda points."[17]

The governors of New Jersey and New York, responding to the public outrage over the Soviets' massacre, made it clear they would not permit Gromyko's plane to land at New York's Kennedy International Airport or at the Newark, New Jersey, airport.[18] The Soviets seized on this action, claiming that the United States had violated an agreement signed in 1947 guaranteeing and facilitating travel to and from the United Nations for foreign diplomats. The Soviets rejected U.S. offers to use military airfields for Gromyko's arrival.

New York and New Jersey's actions against the Soviets were not isolated; nineteen states removed imported Russian vodka from their state-owned and -run liquor stores. The move had started with Ohio which, frustrated at the Reagan administration's lack of action, took the lead. "The state of Ohio," wrote columnist George Will, "which has a better foreign policy than the United States, has removed Russian vodka from state-run liquor stores. Perhaps the 269 murders will complicate the process of subordinating foreign policy to presidential politics."[19]

On September 19 the Reagan White House and State Department strategy of saying much but doing little nearly blew up in their faces. In a heated debate at the United Nations with the Soviet delegate, Deputy Ambassador Lichenstein casually suggested, in responding to the Kremlin envoy's charge that the United States failed to accommodate Gromyko, that U.N. members who did not feel welcome could consider moving the international forum to another country. "We will put no impediment in your way," Lichenstein added, "and we will be at dockside bidding you a farewell as you set off into the sunset."[20]

The statement was like throwing a match into a powder magazine, setting off an explosive public reaction and outpouring of public support for such an action. Telephone calls and telegrams poured into the U.S. Mission to the United Nations in New York running thirty to one in favor! If it was a vote of confidence for the fifty-seven-year-old former CIA agent, it was conversely a vote of no-confidence for the Reagan administration's strategy of taking its case to the United Nations. In fact, outspoken Lichenstein had predicted just prior to the Soviet veto of the watered-down Security Council resolution: "Nothing tangible will result from this Council vote. Does the U.N. ever do *anything*?"[21] Lichenstein's incendiary comment was not an "off the cuff" remark as administration officials contended when they ordered him to be put on ice while issuing a statement maintaining that his views did not represent a change in policy.[22] In fact, Lichenstein had submitted his resignation to President Reagan *prior* to making the remark because of the frustration encountered at his U.N. post.[23]

President Reagan may have momentarily shared the same frustration. On September 21 Mr. Reagan suggested that perhaps U.N. delegates would have a valuable comparison if they had to spend six months in Moscow and six months in New York attending sessions. "I think the gentleman who spoke the other day," the president said of Ambassador Lichenstein, "had the hearty approval of most people in America in his suggestion that we weren't asking anyone to leave, but if they chose to leave, goodbye."[24]

Nevertheless, on September 22, the following day, the Reagan administration and liberal Republicans were objecting to Senate passage, by a lopsided sixty-six to twenty-three vote, of a proposal that would have slashed U.S. contributions to the United Nations by $480 million over a four-year period. The sponsor of the surprise measure and a supporter of the U.N., Senator Nancy Kassenbaum (R–Kan.), insisted that her proposal was aimed at cutting the fat out of the international organization. Other senators voted for the proposal over White House and State Department objections as a way of sending a message to the Soviet-dominated United Nations.[25]

In fact, a study published shortly after the Senate vote revealed that top officials at the international organization received more in salaries than Ambassador Kirkpatrick! While she received an annual salary of $69,600, the U.N. Secretary General was paid $139,300 annually, twenty-eight undersecretaries received $96,765 each, and twenty-three assistant secretaries took home $85,864 annually.[26]

Two days before the president's speech at the U.N. General Assembly, the Reagan administration, reported the *Post*, "in a change of tactics," rushed to the defense of the United Nations to try to overturn the Senate vote.[27] The failed strategy of the White House at the United Nations and the failure of the organization to accord a minimal amount of justice to the murder of 269 people were ignored. The issue became defense of the United Nations, ignoring the tragedy of KAL 007 and a long list of other issues.

In his U.N. speech, President Reagan made only a passing reference to the airline massacre, devoting the bulk of his speech to disarmament proposals and praise of the world organization. He did, however, ask what had happened to the dream of the U.N. founders in 1945, answering his own rhetorical question by asserting that governments had gotten in the way. The international organization was founded to speak with moral authority, Mr. Reagan said, "that was to be its greatest power," but "somewhere the truth was lost that people don't make war, governments do."[28]

Ambassador Kirkpatrick, a week after the president's U.N.

address and after the Soviets did their bashing in of his arms reduction proposals, spoke before the Heritage Foundation in Washington, avoiding any reference to the shipwreck of the administration's strategy at the United Nations over KAL 007. Instead, she concentrated on the scope of Soviet global ambitions, the product of a "tradition of Oriental despotism" that relies on lies and violence. "The neglect of history lies, I believe, at the root of most of our foreign policy failures," she said, citing the situations in Iran and Nicaragua as having long been known to the experts before internal upheavals occurred.[29]

Burton Pines, former *Time* magazine editor and vice president of the Heritage Foundation, would not quarrel with Mrs. Kirkpatrick's basic thesis. However, he is in a unique position to judge the record of the United Nations, having supervised a massive study of it undertaken by Heritage. Appearing on ABC-TV when the momentary public furor erupted over whether the United Nations should go elsewhere, he made the point that the United Nations has, among many things, given legitimacy to terrorist groups like the Palestine Liberation Organization. Pines stated:

> I think it would be a marvelous idea if the United Nations were moved to Moscow. I think it would be very good, first of all, for the United States to have the same opportunities for espionage in Moscow that the Soviets have enjoyed here for three and a half decades. And it would also be marvelous that all those delegates from the Third World countries who romanticize the marvelous life in the Soviet Union would confront what life really is like in a totalitarian society.
>
> The Heritage Foundation studies are only half way through its time frame. But if you pushed me to the wall, I would say that, so far, evidence seems to indicate the United Nations is not serving peace—or almost any of the purposes for which it was founded. It has not kept the peace. It does not make the peace. The United Nations increases tensions rather than lessens tensions. It politicizes issues which need not be politicized. It globalizes regional issues. It imposes an enormous cost on the United States, not the $200 million a reporter mentioned before; but it currently costs the United States one billion dollars every year for the U.N.[30]

Congressman Lawrence McDonald had made similar arguments over the years. In September of 1979 a national news magazine asked the Georgia Democrat whether the United Nations had been a success or a failure. He replied it was a dream that not only had failed but also had turned into a nightmare for the United States and the free world.

> Under the auspices of the United Nations we have seen bloody wars in Korea and Vietnam, the subjugation of Eastern Europe and the mass slaughter of tens of millions of people in China. We have witnessed the Soviet invasions of Hungary and Czechoslovakia. We are witnessing today the genocide in Cambodia. In all of these instances, the United Nations did nothing. . . . I think we should admit reality. The U.N. has become a smokescreen for dilution of American sovereignty, for the diverting of American purpose, and the whittling down of our potential for world leadership. It is also now and has been a cover for worldwide Communist aggression.[31]

18

The Crusade of the Craven

***"If we had not been lying to ourselves about the nature of the Soviet dictatorship, we would not have been shocked at the murder of 269 people."*[1]**

U.S. Rep. Newt Gingrich (R–Ga.)
September 14, 1983

When KAL 007 was destroyed, the U.S. House of Representatives was in summer recess and did not reconvene until September 12. House leaders and members, therefore, had almost twelve full days to familiarize themselves with the facts and to gauge the mood of America. In the immediate aftermath of the tragedy, House Speaker Thomas "Tip" O'Neill (D–Mass.) perhaps spoke for most Americans when he called the massacre "an unbelievably barbaric act" of airborne terrorism that will "not be forgotten or excused by decent people over the world."[2]

When the House, however, reconvened and took up consideration of Joint Resolution 353 on September 14, the record clearly establishes that Speaker O'Neill and the House (with very few exceptions) were united in their purpose to do nothing that would force the Reagan administration to take concerted action against the Soviets. The late Congressman Clement J. Zablocki (D–Wis.), chairman of the House Foreign Affairs Committee, said:

> It is my hope that this joint resolution will contribute to the international effort initiated by the president to hold the Soviet Union responsible for this criminal act as well as to develop improvements in civil aviation procedures to prevent its recurrence.[3]

The specific eight points in the resolution called for no sanctions against the Soviets, although the ranking GOP member of the House Foreign Affairs Committee, Congressman William S. Broomfield (R–Mich.) immediately followed Zablocki with the proposal that "our Government expel 269 Soviet KGB agents, one for each of the victims of the Korean Airlines Flight 007. I think it would be a symbolic act that would have a practical impact on the Soviets' worldwide espionage operation."[4]

Broomfield did not propose that his action item be made an amendment to House Joint Resolution (HJR) 353. It was clear that Congressman Zablocki, House floor leader for the majority Democrats, would not permit or even consider any amendments.

Congressman Henry J. Hyde (R–Ill.) candidly observed that since Congress had learned to live with past Soviet barbarism and tolerated Soviet-sponsored terrorism in Central American nations like Nicaragua, it was therefore reasonable to assume that the resolution would follow the same pattern. "This is probably a useful thing," Hyde told the House, "although this resolution is, as we would say in my old neighborhood, all windup and no pitch. It views with alarm and it condemns and it deplores but it really does not do anything."[5]

Remarkably, neither Zablocki nor Broomfield nor Hyde mentioned in his remarks the all-too-obvious fact that one of their House colleagues, Lawrence P. McDonald, had perished on KAL 007. During the entire debate on HJR 353 (which did mention the Georgia Democrat by name), over ninety Democratic and Republican House members spoke, but *only fifteen* mentioned McDonald by name. All but two mentioned him in passing, and an additional three referred to him only as "our colleague," with rare reference to his repeated warnings to the House about the nature of the Soviet threat.

In fact, the first in the House debate to focus in on this very fact was the nonvoting member of the House from Guam! "What an irony," Antonio Borja Won Pat told the House, "that our colleague, Larry McDonald, should have died in such an incident. His warnings that the Soviets were barbarians ring truer now than ever before."[6]

As with President Reagan's inability to bring himself to mention McDonald by name, a majority of House members exhibited an embarrassing lack of public comment over the murder of one of their own, while appearing to evade the ironic significance of his destruction and that of 268 others. (Several congressmen and a congresswoman mentioned, rightfully so, the names of constituents who had perished on KAL 007, but they did not mention their former colleague.)

Congressman Thomas J. Bliley (R–Va.), one of only three House members with a sense of history, reminded the House that KAL 007's destruction by the Soviets was not some isolated incident. Wholesale murder followed the Russian revolution and civil war after 1917, Bliley said, while Soviet dictator Stalin killed up to 10 million of his own people in the forced collectivization of Russian farms in the 1930s. Stalin's annexation of Latvia, Lithuania, Estonia, and Belorussia involved mass murder, only to have him conclude a defense pact with Hitler in 1939.

Bliley cited that in the post-World War II period Stalin overran Eastern Europe by wholesale murder. His heirs snuffed out the liberty and the lives of people in Hungary, Czechoslovakia, Poland, and Afghanistan. The Virginia Republican added:

> People in the West are shocked, outraged, and surprised. I, too, am shocked to my very core. I, too, am outraged to my very core. I am not surprised. This is exactly what our late colleague from Georgia tried to tell us while he lived. Now he is dead. He is dead at the hands of the very government that he tried to expose.[7]

The reaction of the House members to McDonald's death contrasted sharply with the great favorable comment and sympathy they had shown when Congressman Leo Ryan (D–Calif.) was murdered in Jonestown, Guyana, in 1978. Ryan, a liberal, had courageously gone to Guyana to investigate the activities of socialist cult leader Jim Jones, only to be killed by some of his supporters shortly before the murder-suicide of over nine hundred of his followers. The House held public hearings on Ryan's murder, and he was hailed in both the House and the news media as a heroic martyr. None of this type of accolade was accorded McDonald.

Why?

It fell to McDonald's friend and fellow Georgian, Congressman Newt Gingrich, to underscore the real reasons for the remarkable lack of comment by a majority of House members of both parties over the first murder of a U.S. congressman by a foreign power:

> If Larry McDonald had stood on the floor of this House and warned that the Soviets would deliberately, coldly, and ruthlessly murder 269 men, women, and children, many members of this House and of the national news media would have called him paranoid; indeed, many did call him paranoid. Well, it was not paranoia that killed Larry McDonald; it was a deliberate action of the Soviet leadership.
>
> Even today, during this debate, we have heard calls for avoiding any increased military preparedness, for trying to help the Soviets adjust to reality. It is we in the United States who must adjust to reality. It is we who must rethink our interpretations of the last decade.[8]

Congressman Gingrich, a Republican, also pointed out that McDonald was a loner, a Democrat "who refused to vote for the Speaker of his own party, a strong conservative who marched to the beat of a drum that others did not hear. Many people thought he was strange; almost everyone knew he was not typical."[9]

House Speaker O'Neill and the Democratic leadership had always found the medical physician–turned Democratic House member a disruptive influence in their drive for party loyalty. McDonald, however, viewed the House leadership as hypocritically intolerant, who sought to punish him for his dissent by denying him committee assignments that the House seniority system entitled him to.

McDonald had made this an open issue in January 1983, insisting that "party loyalty" was a smokescreen to obscure the fact that the national Democratic leadership had been taken over by "radical liberals" who did not represent the broad mainstream of Democratic thought in the United States, but rather radical special interests who were hungry for power at the expense of the U.S. Constitution.

"I don't vote 'with' any party in the House. I vote with," McDonald said, "the U.S. Constitution and according to the views of the people I represent in northwest Georgia."[10]

But larger issues lay hidden beneath the House leadership's remarkable silence and lack of strong initiatives to punish the Soviets for the destruction of KAL 007. Ever since the election of Ronald Reagan in 1980, the Democratic House leadership in particular and the national Democratic party in general had made the president out to be a moral monster when it came to the poor and the minorities at home, and a trigger-happy Nuclear Napoleon abroad, ready to shoot it out with the Soviets. The Reagan administration, as has been documented, demonstrated by its response to KAL 007 that this portrait was a facile falsehood. The White House and State Department followed a course of appeasement of the Soviets that the liberal Democrats and the House leadership had themselves advocated ever since Mr. Reagan became president.

Speaker O'Neill and the liberal Democrats could not now be expected to advocate a stern policy toward the Soviets; and they certainly could not be expected to shed any tears or publicly eulogize McDonald, who had openly denounced that very same leadership as power hungry at home and as appeasers of the powerful Soviets abroad. The cold and callous mind-set of the liberal House members was also illustrated only a month after the destruction of KAL 007 at a Democratic party fund-raising dinner.

According to a House member who was present, but who requested of this author that his name not be used, Democratic national chairman Charles Manatt, who claimed his was the "party of compassion" as compared to the Republican party of the cruel and the rich, called on Congressman William Gray III (D–Pa.) to lead the invocation at the start of the dinner. Congressman Gray, a black Baptist minister from Philadelphia, bitterly resented McDonald's open and lonely opposition to a proposal for a national holiday in memoriam of slain civil rights leader Dr. Martin Luther King, Jr.

> He asked everybody to pause for a minute of silent prayer in tribute to Senator Henry Jackson of Washington, who had died of natural causes on the same day McDonald was murdered by the Soviets. He never mentioned Larry, and many of us just could not believe it and were deeply offended and angry that a professed Christian and an ordained minister could not seem to rise above his personal political disagreements to mention McDonald and ask that we all pray for him.[11]

The incident recalls McDonald's own words when he openly took issue with the House leadership's denial of committee assignments. Invoking George Washington's warning in his Farewell Address of the dangers arising from a small political elite demanding loyalty to party over country, the Georgia Democrat said he took an oath to uphold and defend the Constitution, not to uphold a particular organization, church, lodge, person, or a radical liberal dogma espoused by a few who, by the accident of numbers, held power in Washington.

In January 1983 McDonald said:

> While radical liberals speak of tolerance, they are intolerant. While liberals preach love, they practice hatred. While radical liberals proclaim their faith in Democracy, they exercise totalitarian rigidity. While liberals give lip service to diversity, they demand adherence to their narrow dogma . . . or else.[12]

During the September 14 House debate over the Joint Resolution, the Democratic House leaders allowed members from both parties to have their angry say, while making it clear they would not allow any proposals that would punish the Soviets. Behind this stern refusal lay cold, calculating politics aimed to preserve their positions.

KAL 007 had provided the Reagan administration a weapon against the House Democrats to force passage of its defense expenditures that had been held hostage prior to the Soviets' shooting down of the airliner. Congressman John F. Seiberling (D–Ohio) expressed the dilemma of the dominant liberal House Democrats when he pleaded that KAL 007 be judged separately from increased defense expenditure requests and funds for the MX missile. Raising the banner of budget limitations, Seiberling best expressed the prevailing liberal interpretation of KAL 007.

> If anything, the Korean Airliner atrocity, by revealing the rigid, even paranoid, Soviet military mentality, emphasizes the precariousness of the situation humanity finds itself in as a result of the nuclear arms race. In these circumstances, the addition of an unnecessary, destabilizing weapon such as MX can only heighten the chances of a nuclear holocaust being brought on by some military or civilian official somewhere overreacting to some perceived threat, just as the Soviets did in the case of flight 007.[13]

Most House liberals, while deploring the destruction of KAL 007, argued for continued arms control talks with the Soviets on the assumption that the airliner's destruction was accidental and not premeditated. In raising the threat of nuclear conflict, they were using the incident of terrorism to try to terrorize the House out of passing (as it eventually did) additional defense expenditures and money for the MX missile. However, both the soft- and hard-liners in the House used KAL 007 for their own purposes. The conservative hard-liners proposed a long list of punitive measures: reducing U.S. funding at the United Nations, using import taxes on Soviet caviar and vodka as a fund for families of the victims, suspending arms control talks, a grain embargo and cut-off of transfer of technology, end to loans and credits to the Soviets and Eastern European Bloc countries, a break in diplomatic relations, banning USSR participation in the 1984 Los Angeles Olympic Games, and the banning of Soviet imports.

None of these proposals was ever formally introduced as part of the HJR 353 because everyone in the House knew that Speaker O'Neill had the votes from the Democrats to stop their passage, while the Republicans knew that to push them would run counter to what the Reagan administration wanted. The Republicans in the House successfully used the KAL 007 massacre to gain passage of the administration's defense requests in exchange for promoting a policy of appeasement of the Soviets as favored by the Democrats.

House Joint Resolution 353 passed by a 416-to-0 vote, with 2 answering "present" and 16 not voting. What emerged was a resolution that condemned the crime and the criminal, but suggested no punishment.

What the nation was presented with was a Crusade of the Craven in Congress, one that repeated a pattern developing for over a decade in the Congress and which Congressman McDonald had fought. He had been elected a few months before the United States withdrew from Vietnam in 1975, abandoning Southeast Asia to the Communists. During almost nine years in Congress he had watched his fellow House members lie to themselves and to the country about the real nature of the menace to the free world.

Congressman Gingrich observed on the House floor:

> For almost ten years we have lied to ourselves about the conquest of Vietnam, the use of poison gas in Cambodia and Laos, the Soviet use of Cuban puppets in Angola and Ethiopia, the establishment of the Soviet military threat in Cuba, the creation of a new pro-Soviet puppet in Nicaragua, and the creation of a KGB terror network that supports violence and disinformation across this planet.
>
> If we had not been lying to ourselves about the nature of the Soviet dictatorship, we would not have been shocked at the murder of 269 people. And that is the simplest of all tests. If you were surprised that they would kill 269 people, you have not been watching them very carefully. . . .
>
> This resolution is not a ceiling above which we cannot be stronger. This resolution is the basement on which we build the actions of the future. If this House and the members of this House vote and forget, if the news media reports and ignores, then Larry McDonald and 268 other people will have died in vain. But if a year from now our defense is stronger, our negotiations are tougher, our awareness is greater, then this resolution will have been the first step toward saving freedom on this planet.[14]

19

A Majority of Chamberlains, a Minority of Churchills

***"A new Munich indeed is stirring in some minds."*[1]**

Sen. Jesse Helms (R–N.C.)
U.S. Senate
September 15, 1983

British Prime Minister Neville Chamberlain's appeasement of Nazi dictator Adolf Hitler at Munich in September 1938 was hailed by just about everyone in the West as a victory for peace, including President Franklin D. Roosevelt.[2] Only later did Chamberlain alone bear the harsh judgment of history as a weak-willed national political leader who groveled before a political gangster like Hitler. However, while Chamberlain was naive, he was not a groveling coward.

Historian Telford Taylor in his definitive study of Munich wrote:

> Chamberlain did what he did at Munich, not because he thought he had to, but because he thought it right. For him, appeasement was a policy not of fear but of common and moral sense. . . . Sadly mistaken he may have been; cowardly or indecisive he was not, and for him Munich was no surrender, but a passionately moral act.[3]

When the U.S. Senate took up consideration of Joint House Resolution 353 on September 15, 1983, Sen. Jesse Helms was virtually alone in publicly voicing the view that the West's response to the KAL 007 incident was reminiscent of the Munich period of appeasement. He told the Senate:

> In the wake of the Korean airline massacre, voices are rising to counsel appeasement. We hear them in the press, on radio, on television, in academic circles and in business circles, in politi-

> cal circles, and in legislative bodies throughout the West. The Soviet Union, however, continues with relative impunity to impose its design of terror and expansion on the world.[4]

Helms had led a hard core of conservative senators in a futile effort to secure passage of eight amendments to the House Resolution that would have advised the Reagan administration on what specific actions could be taken against the Soviets. Those proposed actions were:

1. Recall of the U.S. ambassador to the USSR for consultations, and reduction of the number of Soviet diplomatic personnel in the United States to the number accredited to Washington by Moscow.
2. Comprehensive reappraisal of U.S.–Soviet relations, including arms control, human rights, East-West trade, and regional issues.
3. Report by the Reagan administration to the Congress on the compliance or lack of compliance of the Soviets with the spirit and letter of all existing strategic arms limitation treaties.
4. Linkage of future arms reduction talks with Soviet willingness to abide by international laws, with specific attention to KAL 007, Soviet violations of the Helsinki Human Rights Accords, its invasion of Afghanistan, its repression in Poland, and use of chemical and biological weapons in violation of existing treaties.
5. Reemphasis of the Monroe Doctrine in the Western Hemisphere due to the presence of Soviet troops in Cuba.
6. Declaration that Poland was in default on all or part of its debt to the West, and acknowledgment that credit to Communist nations is an element of national strategy.
7. Tighten substantially the control over the export to the Soviets of tools, high-technology products, and equipment for the development of Soviet oil and gas resources.
8. Use of existing law to prevent the import from the USSR to the United States any product or material produced by forced labor.[5]

"The administration strongly opposes the Helms amendments," Sen. Charles Percy (R–Ill.) told the Senate. The chairman of the Senate Committee on Foreign Relations then read into the record a memo authored by the administration, setting forth its reasons. It was the first time during the discussions that anyone was able to get a fix on the White House and State Department's thinking behind their refusal to consider punitive measures against the Kremlin.[6] (See Appendix II for the administration's memorandum to Sen. Charles Percy.)

Percy, as Senate floor manager, urged defeat of the Helms proposals, reinforcing the fear, as expressed in the House and in the national news media, that without business as usual with the Soviets the potential of nuclear war between the superpowers would be increased. Percy argued:

> We can only imagine, and we now have a basis for fearing—the whole world has—how the Soviets might react if they thought their radar showed they were under a nuclear missile attack. . . .
>
> We want to confirm the House action today, which was to show the world that the president, the House of Representatives, and the Senate stand united with the same words with one voice in condemning the Soviet Union for its wanton act against international civil aviation and against humanity itself.[7]

Thus, the entire U.S. Congress and the Reagan administration are on record as responding to a specific act of Soviet terrorism out of fear of what Moscow *might* do if the proposals by Senator Helms were adopted! Senate majority leader Howard Baker (R–Tenn.) tried but failed to prevent any amendments from being considered, arguing that the "unity of the purpose" to speak with one voice was more important than making the Soviets pay a price for mass murder.

Senator Baker also made it clear that both the White House and the State Department favored this course of action. He revealed that it was the administration that drafted, in consultation with congressional leaders of both parties, the language of the House and Senate resolutions, and "the clearance process, so called, extended to representatives of the State Department, the professional Foreign Service, the White House, and the National Security Advisor's office."[8]

In other words, the resolutions had passed through many hands, and *not one of those thought it proper during the drafting stages to include the name of Congressman Lawrence McDonald.*

Senator Baker inadvertently revealed that it was not until this omission by administration drafters was noted by the chairman of the House Foreign Relations Committee, Congressman Clement Zablocki (D–Wis.), that McDonald's name was included.[9]

The omission, the drive to prevent any amendments, and the effort to rush the resolution through the Senate provided clear evidence that Congress and the Reagan administration, with support from the national news media, wanted to be quickly done with KAL 007 so that they could get back to business as usual. Ignoring the indignant anger of the average American about the Soviet crime's going unpunished, the bipartisan appeasement policy was cloaked in the fear of nuclear war and then surrounded with an alluring lie that the issue was the world versus the Soviets.

Senator Percy told the Senate:

> It is clearly now the strategy of the administration, because it is in accordance with the feelings of the people of the world, that it is not just a U.S. issue. It is not the United States versus the USSR. It is not an East-West confrontation. It is the world, the whole world, that feels revulsion about this action.[10]

However, unlike the leaders in the House, the Senate GOP leadership could not prevent the record from being made by Senator Helms and others that the KAL 007 massacre was a warning to the West. Senator Steve Symms told the Senate:

> The Soviets, it is safe to say, have never been truly set back by any of the free world's reactions to their incredible history of aggression and relentless pursuit of their longstanding geopolitical goals. Their well-founded belief is that emotions in the West are short-lived and that past policies of appeasement will resume. The president's outrage must be met with appropriate action. We must demonstrate that barbarism has consequences more damaging than mere bad publicity.[11]

The national news media sought to represent Helms and Symms as a tiny minority favoring strong action against the

Soviets. The eight separately recorded votes in the Senate to "table" (meaning to kill) Helms's proposals tell an entirely different and revealing story.

On the first two votes, twenty-five and twenty-six senators respectively voted with Senator Helms.[12] On the third vote to table, forty-five voted nay (*not* to kill the amendment), while on the fourth vote only fourteen voted with Helms.[13] On the sixth vote twenty-nine sided with Helms and on the seventh tabling motion (to kill the amendment), forty-three voted with the North Carolina Republican.[14] On the eighth and final amendment, *forty-five* senators sided with Helms.[15] Thus, the record clearly shows that almost half of the Senate favored some form of specific action against the Soviets, while the Republican Senate leadership and the Reagan administration wanted none!

Senator Robert Byrd (D–W.Va.), the Democratic minority leader, was also intent on making it clear just where the administration stood. After the first seven votes to table Helms's amendments, Senator Byrd asked majority leader Baker whether it was not in fact true that the White House favored the tabling action? Baker replied:

> Just before this series of votes began, I reconfirmed that with the president's principal staff and, indeed, the notation on my advisory calendar here during the vote showed that the administration, the leadership on this side, the jurisdictional committees, supported tabling in each instance.[16]

Senator Byrd also made part of the Senate record that Sen. Alan Cranston (D–Calif.) and Sen. Ernest Hollings (D–S.C.) were not present for any of the eight votes, nor were they or Sen. John Glenn (D–Ohio) present for the final vote and passage by ninety-five-to-zero of the Senate Resolution condemning Soviet destruction of KAL 007.[17]*

Senator Glenn, along with senators Cranston and Hollings, was an announced 1984 Democratic presidential candidate.

---

*Senators who did not vote were: Alan Cranston (D–Calif.); John Glenn (D–Ohio); Ernest F. Hollings (D–S.C.); Russell B. Long (D–La.); and Larry Pressler (R-S.Dak.).

The Ohio Democrat and former U.S. Marine officer and astronaut voted for tabling four of the Helms proposals, voted with Helms on one, and was absent on three of the votes, not including the final one for passage. In fact, the Senate debate record for September 15, 1983, reveals no statement by Senator Glenn on the KAL 007 incident.

Senator Edward Kennedy (D–Mass.) spoke for the Senate liberals when he pleaded the temporary insanity argument by calling the Soviet destruction of KAL 007 a "single moment of international madness." He also joined liberal Republicans like Senator Percy in using the fear of nuclear war as the basis for doing nothing. Senator Kennedy told the Senate:

> I believe that the president also deserves credit for resisting the irresponsible right-wing pressures to use this tragedy to block progress on arms control. While this latest act of Soviet brutality makes arms control negotiations more difficult, it also makes them more urgent and vital.[18]

Senator Claiborne Pell (D–R.I.) was chairman of a nine-member Senate delegation who had met with Yuri Andropov for two hours in Moscow in late August 1983. Senator Pell told the Senate he came away "cautiously optimistic" about a new Soviet arms proposal, only to return to Washington and find out within hours that the Kremlin had sanctioned the destruction of 269 innocent lives. Pell described the act as the product of "incompetence" and said the Kremlin's "insensitivity to international opinion is deeply worrisome, and, in the setback to East-West relations, the entire world has lost greatly."[19]

Senator Jake Garn (R–Utah) observed that earlier in 1983 Andropov was described in the Western news media as a "closet liberal," but he had shown his true colors by the downing of the airliner while using the Big Lie technique perfected by Adolf Hitler. The Soviets' lying and "stonewalling the truth over the KAL 007 massacre," Garn added, were fully consistent with the brutal background of Andropov as head of the Soviet KGB and with the terrorizing heritage of the Kremlin dating back to Lenin, Stalin, and their successors. "I would suggest," Garn told the Senate, "that the magnitude of the Soviet Union's crime against humanity could be exceeded only by the crime we could commit by forgetting it, and ignoring what it tells us about Soviet values."[20]

Senator Daniel Patrick Moynihan (D–N.Y.) had formed a partnership with Republicans and the Reagan administration, as did many other Senate Democrats, in fostering the idea that branding the Soviets as international criminals was somehow an effective response to the destruction of KAL 007. Over and over again in the Senate debate Moynihan made the point that never before in the history of Congress has a resolution condemned the Soviets as criminal, adding:

> This is not a small thing; there has never before been such an event. The charge of crime under international law is a solemn one. It was not thought to exist. . . . I hope this measure would go forward from this Chamber with the understanding by the president and the secretary of state that when we declare an act to be criminal, we expect the Executive Branch to pursue the matter.[21]

Aleksandr Solzhenitsyn has pointed out that in the Soviet Union the concept of law is nonexistent—a fact made obvious by the Soviets' conduct. As the former Moscow correspondent for the *New York Times*, Hedrick Smith, wrote in his best-selling book, *The Russians*, "what matters most is power. The Russian obeys power, not the law."[22]

It is a profound mistake to assume that the advocates of the bipartisan appeasement of the Soviets over KAL 007 are products merely of weakness and evil motives. They wrongly assume—as did Chamberlain and the British liberal establishment of the 1930s—that the sole choice is between war and peace. Free societies are always faced with the problem of how to cope with totalitarianism and still survive with their liberties in place, rather than bartering them away piecemeal for momentary safety and security.

The historian Telford Taylor, in the closing pages of his definitive work on Munich, quotes Aleksandr Solzhenitsyn from his 1972 Nobel Prize-winning lecture.

> The spirit of Munich is not a thing of the past; it was more than a short episode. I would even venture to say that the spirit of

Munich is predominant in the twentieth century. The entire civilized world trembled as snarling barbarism suddenly re-emerged and moved into the attack. It found it had nothing to fight with but smiles and concessions. The spirit of Munich is an illness of willpower of rich people. It is the everyday state of those who have given in to the desire for well being at any price, to material prosperity as the main aim of life on the earth.[23]

20

Mass Murder and Moral Twilight

> ***"If we are willing to accept in commerce the goods produced in the Gulags . . . we are, in a sense, accomplices to the Gulags themselves."***[1]
>
> Sen. William Armstrong (R–Colo.)
> October 26, 1983

On September 15, 1983, when the U.S. Senate passed the Joint Congressional Resolution condemning the Soviet destruction of KAL 007, it was sent immediately to the White House for President Reagan's signature. Releasing a statement later, Mr. Reagan said, "I salute [the House and Senate] for their overwhelming, bipartisan support," insisting that the Kremlin was now on notice that as Americans the nation was "united and determined to protect our freedom and secure the peace." The president added:

> I urge the American people to stand by the families whose loved ones were taken from them. And let us appeal to the conscience of the civilized world: The passengers of KAL 007 must never be forgotten; nor must we rest until the world can prevent such a crime against humanity from ever happening again.[2]

When it became clear that such words represented an effort at tranquilizing the rage of a majority of Americans, the observation made the rounds in Washington that Mr. Reagan demonstrated more toughness when the air traffic controllers went on an illegal labor strike than he did with the Soviets over KAL 007.

The action that the president did take had no real impact on the Soviets. The closing of the two Soviet Aeroflot airline offices in the United States did nothing to injure the Kremlin economically, particularly since Aeroflot landings in the United States had been banned since January 1982, after the Warsaw-Moscow axis imposed martial law on Poland. The Aeroflot offices, until they were closed by Mr. Reagan, had been devoted mainly to booking passengers on flights that landed in Canada and other countries.[3]

Canada had been the first country to ban Soviet passenger planes from landing on its soil for a period of sixty days, although the ban did not cover completely Soviet-Cuban aircraft using its Gander, Newfoundland, facilities for refueling of regular Soviet flights from Moscow to Havana.[4] While almost all of our Western allies were as rhetorically fierce in their condemnation of the Soviets as the United States, the record shows that the free world was as unwilling to take punitive measures as was the United States.

In fact, it was only through angry protests by the world's airline pilots that many governments in the West went along with a proposed boycott of flights to Moscow, while refusing to ban through the West all Soviet flights to their capitals. "Resolute action," reported William H. Gregory of *Aviation Week & Space Technology,* "by pilots at the working level forced governments in the end to go along, even though civil servants clearly would have preferred to pussy-foot to avoid antagonizing the Soviets."[5]

The *Washington Post*'s Paris correspondent, Michael Dobbs, reported that in Western Europe there was little support for economic sanctions, halting technology sales to the Soviets, or suspending arms talks. Dobbs wrote:

> Foreign ministry spokesmen contacted by the *Washington Post* correspondents in several Western European capitals said they were waiting to hear President Reagan's speech tonight [September 5, 1983] before deciding what measures to take themselves.[6]

When Mr. Reagan made it clear that he would confine his response to cultural and aviation activities, any momentum for tough action from the Western allies was wiped out.

The action of the International Airline Pilots Association,

numbering sixty-one thousand members, is one illustration. On September 7, 1983, in London the association's president, Captain Robert Tweedy, announced that its member pilots would support a sixty-day boycott of selected flights to Moscow.[7] On September 12, 1983, the boycott went into effect.[8] Eighteen days later the same association called for early suspension of the sixty-day ban!

The association president told reporters that the sudden reversal was the outgrowth of moves by the international community to improve the rules of the air, and of Soviet initiatives and pledges for more stringent procedures for the interception of off-course civilian airliners by military aircraft.

Captain Thomas Ashwood, the first vice-president of the U.S. main pilots' union, who flies for Trans World Airlines, said Soviet "flexibility" had been detected in recent Kremlin public statements, including the admission that KAL 007 had been shot down by mistake and that the USSR military had been wrong in not promptly admitting what had happened. "There have been indications from the technical aviation sources that the Soviet Union is willing to cooperate in methods to prevent a recurrence of such an event," Captain Ashwood said.[9]

It was hard to believe that airline professionals could accept such Soviet pledges, given the gross degree of lying that the Kremlin exhibited in the first two weeks of the KAL 007 crisis. The airline federation officers had earlier rejected as "absurd" the Kremlin cover story that the 747 was a spy plane. Captain John LeRoy, an American pilot and an officer of the federation, maintained:

> The Russians on the ground knew exactly what it was: a commercial airliner on a routine flight. I have no doubt that they knew what it was. But no one ever thought that a 747 filled with people would be shot down. No one thought that anyone would be that inhuman.[10]

Just prior to the decision to abort the boycott, stories began surfacing in the press quoting an advisor to the Reagan administration on tourism, Cord D. Hansen-Sturm, that actions against the Soviets were ineffective and that the Soviet airline

Aeroflot was grossing $2 million in additional revenue per week because of the boycott—because the Soviets had put on additional flights from Europe to the Soviet Union. "A trade sanction that prevents thinking people instead of inert goods from crossing borders is like shooting ourselves in the brain," Hansen-Sturm said in a report to the Office of the U.S. Trade Representative headed by William Brock.[11]

The Reagan administration voiced no objection when the Airline Pilots Association ended its boycott action, raising the question of whether Brock and others in the administration had pressed for its early end. There is no evidence on which to base a firm conclusion that this in fact was the case. But there is evidence to demonstrate that in the wake of the toothless House-Senate Joint Resolution, the Reagan administration persisted with a policy that resisted almost all efforts, inside and outside the administration, to impose specific penalties on the Kremlin.

On September 19, 1983, for example, a task force made up of various high-level officials of federal agencies was revising earlier moves and recommended that President Reagan deny export licenses to a Hughes Tool Company subsidiary for the sale of $40 million worth of oil exploration equipment to the Soviets on the grounds of foreign policy and national security. The decision to allow the sale had been made by President Reagan roughly ten days before KAL 007 was shot down. Lawrence J. Brady, assistant secretary of commerce and a strong opponent of trade with the Soviets, headed the task force. The *New York Times* quoted an unnamed administration official as saying: "I don't see any way possible that the recommendations will be overturned in view of the president's statements regarding the Korean flight."[12]

Five days later, reports began to surface of a bitter struggle between the Brady group and Secretary of State George Shultz and Secretary of Commerce Malcolm Baldridge, both who favored selling technology to the Soviets. National Security Advisor William Clark was reported as trying to referee the bitter dispute.[13] Within a month after Brady's recommendation to reverse the sale agreement, the hard-liner made it known he was leaving his post to run for elective office in New Hampshire.

New York Times columnist William Safire reported on October 9 that Shultz's State Department objected to the Brady recommendation and that when National Security Advisor William Clark "did not press the hard-line case for fear of being blamed for bagging another Secretary of State, President Reagan meekly acquiesced." Safire concluded:

> The upshot is not merely "business as usual" with the Russians, but "better business than ever." The technology sales are often linked to financing and buy-back arrangements and energy dependency. Rather than restrain our allies in their lust to finance and supply the Soviet Union with oil-gas technology that will enable Moscow to maintain superpower status, Mr. Reagan has decided to compete with the Europeans and the Japanese for Russian business. Not only do we aid the Russians militarily, we strengthen them economically and strategically.[14]

Safire also noted that a month after KAL 007, Mr. Reagan, in a speech before conservative supporters at a Heritage Foundation dinner in Washington, insisted that his administration had slowed the transfer of free world technology to the Soviets! Using the gas pipeline issue as one of several examples how this was not so, Safire maintained that "when it comes to 'the transfer of valuable free world technology' that will *strategically* [sic] benefit the Soviet Union, it was during the Reagan Administration that the floodgates were opened."[15]

Dr. Miles Costick, president of the Washington-based Institute on Strategic Trade, provided a chilling concrete illustration of how Western technology provides enhancement of Soviet military power. According to Dr. Costick his interviews with KGB defectors, who are experts in electronics, revealed that Soviet agents in West Germany gained access to critical technology that led to the Kremlin's duplication of the U.S. Sidewinder air-to-air missile for Soviet interceptor fighters. Dr. Costick told this author:

> The missiles used to destroy the Korean airliner were direct duplications of the U.S. Sidewinder, including the electronic circuitry produced by U.S. firms and sold to West Germany,

> which the Soviets either bought or gained by industrial espionage. We have now been able to establish that the 269 innocent people aboard KAL 007 were sent to their deaths by a Sidewinder type of air-to-air missile fired by that Soviet fighter over Sakhalin Island.[16]

National Security Advisor William P. Clark endorsed the recommendations that had been made to the president by the Brady group. His endorsement was out of the conviction that the continued flow of Western technology to the Soviets would be dangerous to national security.

Clyde H. Farnsworth of the *New York Times* wrote:

> Mr. Clark's position appears to place him in conflict with both Mr. Baldridge and Secretary of State George P. Shultz. Administration sources said Mr. Shultz was furious over the September 13 recommendation [to restrict technology sales].[17]

Within a few weeks, weary of the White House infighting with those who favored business as usual with the Soviets, Clark left his National Security post to replace Interior Secretary James Watt. Rowland Evans and Robert Novak reported that not only was Watt's departure a blow to Mr. Reagan's conservative supporters but that Clark's exit bewildered both domestic conservatives and European allies alike as to what kind of foreign policy Mr. Reagan was conducting. Clark's movement to the Interior Department, added Evans and Novak, gave "a new supremacy for Shultz's State Department working in league" with the soft-liners in the White House "to soften East-West tensions before the 1984 elections."[18]

It is clear that Reagan administration soft-liners, led by Secretary Shultz, were not even willing to offend the Soviets when it came to items produced in the Soviet Union with political or slave labor and imported and sold in the United States. On October 7 for example, at the moment Clark was deciding to move to Interior, the Commissioner of Customs, William von Raab, recommended that three dozen items made in the Soviet Union be banned for U.S. import because they were believed to be manufactured by forced or slave labor.[19]

On October 26 Senator Armstrong initiated a letter signed

by forty-five other senators to Treasury Secretary Donald T. Regan saying that the United States had a moral and legal obligation to enforce a 1930 law that banned the importation of products mined, produced, or manufactured wholly or in part by convict or forced-slave labor. The letter stated:

> This would be true at any time, but the need for enforcement is especially urgent now, in the wake of the Korean Airlines massacre and mounting evidence of increased repression by the Soviet authorities of domestic human rights activists.[20]

Reagan administration officials at the departments of State, Commerce, Treasury, Office of the U.S. Trade Representative, and the Central Intelligence Agency were reportedly out to derail the import ban out of fear that the Soviets would retaliate and make it much more difficult for the United States to collect intelligence. "There's a moral issue here, too," stated Senator Armstrong. "If we are willing to accept in commerce the goods produced in the Gulags, then it seems to me that we are, in a sense, accomplices to the Gulags themselves."[21]

Presidential promises and morality, like the KAL 007 tragedy itself, were lost in the moral twilight that settled over the Reagan administration before, immediately following, and in the weeks after the destruction of KAL 007. In fact, on the day the Soviets downed KAL 007, the Soviets purchased nine hundred thousand metric tons of wheat and corn; and business was brisk in the proceeding weeks when the United States was denouncing Soviet barbarism and making its case in the United Nations and in the international news media.[22]

But there was both a moral and a practical price to be paid for such actions. Labor columnist Victor Riesel reported that the Soviets had so manipulated the Soviet grain deal signed six days before the destruction of KAL 007 that U.S. cargo ships had been frozen out of transporting millions of metric tons, and thus they lost badly needed jobs for the U.S. Merchant Marine. "The civilized world," wrote Riesel, "gets the flotsam of the grim Korean airliner; and the Soviets get the grain at a satisfactory price, good credit rates, and their merchant fleet makes a profit."[23]

One of the last columns that Congressman Lawrence McDonald wrote for distribution in the Seventh Congressional District of Georgia dealt, ironically, with the problem of technology transfer to the Soviets.

> Well over a decade ago conservatives warned that the United States should not provide the expertise necessary to build the huge Kama River truck plant in the Soviet Union.
>
> The Soviets insisted they would use the output from the plant only for civilian purposes—and liberals agreed. Over $1.5 billion of U.S. and Western European technology went into the plant.
>
> But, in the summer, 1977, Kama River trucks were used by Soviet forces in East Germany. They also supported the Soviet invasion of Afghanistan in late 1979 and 1980. And now that they have our technology, the Soviets have announced a new model Kama River truck which will be used exclusively for military purposes.[24]

21

The Prisoner of Pennsylvania Avenue?

"We must have the courage to do what we know is morally right, and this policy of accommodation asks us to accept the greatest possible immorality."[1]

Ronald Reagan
National Television Speech
October 27, 1964

Ronald Reagan is one of the few presidents in this century who pursued a successful career independent of politics and did not spend the majority of his adult life on the public payroll. The key to his success as a Hollywood star was that he had always been able to take direction. Whether a script was good or bad, Reagan could always make it come alive and sound credible.

Robert Walker, former top political advisor to Mr. Reagan when he was governor of California, maintains that others around Reagan ran his political campaigns and administration. Walker observed:

> He never understood the mechanics of the operations that elected him to the governorship twice, and he has no perception of what a national presidential campaign should be like. He'd be a great knight on a white horse; he'd lead the people to victory. But he wouldn't know how to run the government.[2]

Ronald Reagan does not relish face-to-face confrontations that challenge and provoke. The author knows this from two lengthy encounters with him, one in August 1975 and one in July 1979. His advisors since his days as governor of California also know this, as do those whom he appointed to serve him during his first term as president. He prefers to be given direction and performs best when the script has been blocked out. While this has its advantages, it also makes Mr. Reagan appear to be the pawn and prisoner of his advisors in a White House setting that can be as unreal as a Hollywood sound stage.

When KAL 007 was shot down, President Reagan was vacationing on his ranch in California. Secretary of State George P. Shultz was in Washington; he immediately took charge of the crisis, and his direction of the U.S. response dominated from the first day. His influence was decisive, no matter what the president and his advisors might have thought the response should be. In fact, the record clearly indicates that the White House was more concerned about the domestic political ramifications of KAL 007 than they were about whether the response formulated by Secretary Shultz and the State Department was contrary to what President Reagan had previously said or believed.

When this author began this work, he decided that so historic a crisis should have the views of President Reagan and Secretary of State George Shultz, rather than relying on the news media reports and press releases from the White House and State Department. On October 3, 1983, he wrote a formal request for an interview to White House Director of Communications David Gergen. Included with the letter were twenty questions relating to KAL 007.

It is unusual for Washington journalists to submit written questions in advance, although this was done for a specific reason. Because of Mr. Reagan's dislike of surprises and confrontations, specific questions were submitted, informing Mr. Gergen that the president's replies would appear in the appendix of the book in an unedited version. The author even suggested that if the interview could not be done in person, he would accept written responses as an alternative.

On November 1, 1983, Mr. Gergen's deputy, Karna Small, called this author at his Washington office and asked about the treatment of the projected work and particularly how much promotion the publisher planned to commit to the book. Miss Small abruptly terminated our conversation, saying that she had other pressing business, when she was informed that the work would also deal with Congressman Lawrence McDonald.[3] Nothing was ever heard again from either Mrs. Small, Mr. Gergen, or anyone else at the White House.

However, Secretary of State George Shultz replied to the twenty questions submitted to him, and his answers are significant because they represent the first post-KAL 007 responses by a Reagan administration official. (See Appendix I for the full text of Shultz's responses.)

For example, we are told for the first time that the decision to release the radio intercepts was made by the secretary of state, and it was he who decided to hold the extraordinary September 1 press conference in Washington that provided the specific details of how the Soviets had destroyed KAL 007.

"I made this decision," he replied to this author's question, "to have the press conference and to release the information we had, not to inflame public opinion, but because I believe the American people and the world at large needed to know the truth about what the Soviets had done."[4] It was not until the September 5 televised nationwide speech by President Reagan that the actual voice recordings of the Soviet fighter pilots were heard.

Shultz maintained, despite criticism that the release of the radio intercepts compromised intelligence gathering, that it was necessary to show the world that the Soviets had in fact destroyed the 747. Shultz stated:

> It should be remembered that for six days after the attack, the Soviets refused to admit they had shot down the airliner. They said it had left their airspace. They only admitted they had shot it down because we and the Japanese had presented the world irrefutable evidence of the Soviet misdeed. The transcript of the radio transmissions between the Korean airliner and the air traffic controllers at Tokyo's Narita airport provided by the Japanese government clearly indicated that KAL 007's pilot had no indication he was flying off course or that he was being pursued by Soviet interceptors.[5]

During the first few days after the airliner's destruction, widespread fear was expressed by elected political leaders and the news media that the incident would lead to an armed conflict. Others saw the Soviet action as a test of U.S. resolve and an effort to provoke a confrontation. Secretary Shultz stated that he did not regard the incident as a deliberate Kremlin test of U.S. and allied resolve. "At no time was there a concern that the incident would lead to war," Shultz added.[6]

The Reagan administration was criticized for not using the communications hotline between Washington and Moscow. The secretary stated:

> Why the Soviets didn't contact us urgently—over the hotline or any other means—during the two and a half hours they held KAL 007 under observation and claim they thought it was a U.S. "spy plane" is a good question, and one you might put to the Soviets.[7]

During the week following the destruction of the airliner, the news media was filled with numerous stories quoting U.S. intelligence sources that the Soviets had mistakenly shot down the 747 and that the entire affair was a horrible accident. Shultz observed on this point:

> While there is no conclusive evidence whether or not the Soviet pilot knew he was shooting down a passenger aircraft, there is no question that the Soviets deliberately shot down a foreign aircraft without making an adequate effort to warn or identify it. Underscoring the deliberate nature of the Soviet action was Gromyko's subsequent warning that a future border violator would suffer the same fate. Soviet indifference as to whether an intruder is a civilian or military was shown in 1978 when a KAL airliner was fired upon even after the Soviet interceptor pilot identified it as a passenger aircraft.[8]

If the Soviet action was not an effort to test Western resolve, what motivated Moscow to do what most civilized nations thought they would never do? Shultz insisted that the answer must lie in the paranoid character of the Soviets and their obsession with security. "I think this act was an expression of that excessive concern over security," he noted.[9]

Asked whether the U.S. strategy of taking its case to the United Nations failed because of the Soviet veto, Shultz maintained that it succeeded in terms of world condemnation, the two-week boycott of air service, and the "impressive and unprecedented demonstration of international unity which delivered a strong political message to Moscow."[10] With respect to what harm had been done to the Soviets, he replied that "the image the Soviets would like to project of themselves as a peace-loving country has certainly been dealt a severe blow."[11]

Secretary Shultz indicated in his responses the conviction that the policy pursued by the administration was "firm and measured." He stated that established policy, "based on American strength, realism about Soviet aims and motives, and a willingness to talk about matters of mutual concern, has provided the appropriate framework for dealing with this crisis."[12]

"While this incident," he added, "may no longer be on the front pages of the newspapers, it has sunk deeply into the consciousness of the world. It will not soon be forgotten."[13]

Secretary Shultz's responses are significant when viewed in light of the leading role he played in defeating efforts within the Reagan administration and Congress to impose punitive measures against the Soviets. However, President Reagan's endorsement, or tacit approval, of the Shultz approach to the KAL 007 crisis repudiates a series of positions that Mr. Reagan publicly put forth since he first got into politics in the 1960s—the main position being that accommodation of the Soviets was not only evil but immoral!

Ronald Reagan said on television on October 27, 1964:

> The spectre our well-meaning liberal friends refuse to face, is that their policy of accommodation is appeasement, and appeasement does not give you a choice between peace and war, only between fight or surrender. We are told that the problem is too complex for a simple answer. They are wrong. There is no easy answer, but there is a simple answer. We must have the courage to do what we know is morally right, and this policy of accommodation asks us to accept the greatest possible immorality.[14]

In his 1980 acceptance speech at the GOP National Convention in Detroit, Mr. Reagan criticized the accommodation policy of the Carter presidency as weak and vacillating, and he pledged:

> It is the responsibility of the president of the United States, in working for peace, to insure that the safety of our people cannot successfully be threatened by a hostile foreign power. As president, fulfilling that responsibility will be my *number one* priority.[15]

In his inaugural address of January 20, 1981, Mr. Reagan pledged that the enemies of freedom and potential adversaries should not misunderstand that while peace is the highest aspiration of the American people

> we will not surrender for it—now or ever. Our forbearance should never be misunderstood. Our reluctance for conflict should not be misjudged as a failure of will. When action is required to preserve our national security, we will act.[16]

There is no doubt that Mr. Reagan believed what he said at the time he made the above statements. However, the policy endorsed by the president in the aftermath of the KAL 007 massacre repudiated his own words. Like Secretary of State Shultz, Mr. Reagan pledged that the airliner atrocity would not be forgotten. But it was quickly forgotten; and the White House and the State Department encouraged policies that accelerated this process of forgetting so that the nation, the Soviets, and the world could get back to business as usual.

On September 17, 1983, *New York Times* correspondent Steven R. Weisman quoted several White House aides on Mr. Reagan's reelection prospects, predicting that any benefits from his handling of the KAL 007 incident would be short-lived. "We need a win somewhere," Weisman quoted one Reagan aide without naming him, "whether it's Latin America, the Middle East, or with the Russians. If we had just one major foreign policy victory, we'd be in great shape."[17]

On October 25, 1983, President Reagan ordered U.S. Marines, Rangers, and airborne military units to storm the tiny Caribbean island of Grenada, at the request of the eastern Caribbean states, to prevent total domination of the island by the Cubans and Soviets. Also, American students attending a medical university on Grenada, caught in the middle of a government coup but not harmed at the time of the invasion, were brought back to the States with feelings of great relief by both students and their families.

The political effect was to erase completely from the public mind the weak U.S. response over KAL 007 and the deliberate policy of not disturbing the status quo between Washington and Moscow. The U.S. intervention in Grenada also consigned to oblivion the flawed Reagan Middle East policy that led to the truck-bomb murder of over 230 Marines, an act that drew no immediate retaliatory response from the United States.

Grenada allowed the Reagan administration to convey the impression of firmness and tough resolve, while lacking the resolve to deal decisively with the more risky and dangerous problems like the Cubans, the Soviets, Communist satellite nations allied with the Marxist regime in Nicaragua, and the terrorist campaign in El Salvador.

Professor Morton Kaplan, chairman of the international relations department at the University of Chicago, contrasted the Reagan administration's toughness over Grenada to what he termed "the marshmallow approach" to Nicaragua and El Salvador. Three weeks after Grenada Dr. Kaplan wrote:

> We are in the midst of a war for Central America and, if Central America falls, trouble in Mexico will not be far behind. American interests will not be served by a policy so pusillanimous that it cannot take the offensive politically.[18]

Congressman Lawrence McDonald, as early as 1976, began issuing public warnings that Central America in general and Mexico in particular were targets for a Communist take-over. It was for this reason that he opposed surrender of U.S. sovereignty over the Panama Canal by Senate treaty. Three months before his murder, when the warnings he issued became a nightmare reality for Central America, he was critical of the Reagan administration's approach in Central America.

Dr. McDonald described an April 27, 1983, presidential speech on Central America, given before a Joint Session of

Congress, as containing truths the American mass media "had been careful not to tell them" about the activities of the Sandinista government in Nicaragua and the outrages committed by the Communists in El Salvador. Unfortunately, the speech was apparently the work of a committee and, as such, it also contained a needless array of sops presumed to please obstinately partisan Democrats, liberals, or fools whom he lacked the time to disenchant of their gullibility. McDonald added:

> For public consumption, or perhaps because it is true, the Reagan administration responds that we would not dream of throwing out the Sandinistas, that we merely wish to pressure or harass them a little, and/or reduce the flow of arms into El Salvador; that we wish to raise the cost to them of doing what they do, or persuade them to engage in the "negotiations" we seem so desperately to want. It is all such typical State Department cant.

In referring to the Central Americans who are fighting for their countries against the Marxists, Congressman McDonald asked: "How do you motivate men to risk their lives for that sort of goal? How do you get men to risk death, mutilation, or years of miserable captivity in order merely to annoy or inconvenience the enemy?"[19]

22

The Tribunes of Totalitarianism

> ***"Can democratic governments survive the systematic (and unsystematic) distortion of political reality by the press, radio, and television under conditions of mass media?"[1]***
>
> Jeane Kirkpatrick
> U.S. Ambassador to the U.N.
> October 25, 1983

"Newspeak" and "Doublethink" were two terms used by the British writer George Orwell in his classic novel *Nineteen Eighty-Four* to describe the mechanism of totalitarian thought control used by the dictator Big Brother. Neither of the terms was a literary device for the novel but was based on what Orwell had perceived in real life among Western intellectuals.

Orwell's greatest concern was the acceptance of a disbelief, by intellectuals and political leaders with a totalitarian turn of mind, in the very existence of objective truth, that is, that reality is knowable. In January 1946 Orwell wrote:

> The friends of totalitarianism in this country [Great Britain] usually tend to argue since absolute truth is unattainable, a big lie is no worse than a little lie. It is pointed out that all historical records are biased and inaccurate, or, on the other hand, that modern physics has proved that what seems to us the real world is an illusion, so that to believe in the evidence of one's senses is simply vulgar philistinism.[2]

Western news media coverage of the Soviets' destruction of KAL 007 contained heavy doses of Doublethink—the holding of two contradictory thoughts in the mind at the same time—by asking readers, listeners, and viewers to deny the objective reality of what the evidence told them about the nature of the Soviets.

Twenty-four hours after the KAL 007 tragedy, for example, a *Washington Post* editorial demonstrated the editors' clear preference for denying the reality of past Soviet behavior by insisting there were a hundred reasons why a Soviet fighter *"should not"* have shot down the 747 and "no good, explicable reason *why it should have"* (author's emphasis). "No conclusive answer," the *Post* added, "is likely to come soon, if at all."[3]

The *New York Times* displayed a similar willingness to deny past evidence when it wrote editorially that "there is no conceivable excuse for any nation shooting down a harmless airliner," while warning, "Any effort to justify such brutality will surely affect America's judgment of the man newly in charge of the world's other doomsday machine."[4]

Thus, from the very first hours two concurrent themes surfaced both in the editorial comment and in the news coverage of KAL 007: the denial of past Soviet behavior as a basis for judging the Kremlin's action *and* fear of future reaction over the incident itself. This was followed in turn by the media's finding a series of reasons to excuse or explain the actions of the Soviets as accidental, the product of paranoia, or somehow the fault of the United States!

Meg Greenfield, *Newsweek* columnist, was one of the few exceptions, noting that for at least two decades many in America have denied the clear and consistent record of Soviet behavior.

> When a reasonably intelligent adult has known someone, let us say a disagreeable neighbor, for several decades, each new incident or depredation does not come to him as an absolute surprise prompting a revision of opinion and leaving the person at a loss as to what might or ought to be done by way of response. We should be able to deal with the fact of the downed Korean airliner in the context of our experience of several decades of turbulent coexistence with the Soviets, as yet another point in the continuum. It is insane that we should have relapsed anew into one of the first-principles, schoolboy debates over whether their fundamental nature is evil. What does it matter? They are who they are and they have been behaving the way they do for

> years, and the question is, what we are going to do about it—not how many angels (or devils) can dance on the point of a pin.[5]

Meg Greenfield predicted accurately that within five weeks the KAL 007 massacre would vanish from public view, as had the Soviet invasion of Afghanistan within a very short time, because of the widespread and dangerous Western assumption that the Soviet character is susceptible to change. To believe otherwise would leave the advocates of accommodation with the Kremlin without any moral justification for persisting with their policy.

In the first ten days of the KAL 007 atrocity, a consensus developed among liberal columnists, editorial writers, and the national news media as a whole that the massacre was all a horrible *mistake.* A *New York Times* editorial stated:

> The Russians made a grievous mistake that they found too humiliating to confess. But they do not routinely massacre innocent travelers. They compounded the error with denials and countercharges that they have been unable to sustain. But if not browbeaten mercilessly they will recognize the importance of safe transit and channels of reliable communication with American leaders.[6]

Such a view is not new; it has persisted over several decades. The reasons for it are not out of ideological affinity with Soviet and Marxist-Leninist philosophy but instead are rooted in a moral relativist view that denies the validity of evidence and experience, a persistent mental bigotry toward drawing conclusions based on brute facts. What such a denial process suggests is a moral and an intellectual cowardice rooted in fear.

The historian Otto J. Scott suggested that fear was at the root of the national news pundits' one-sided gratuitous attitude toward the Soviets' destruction of KAL 007. Scott told this author:

> I recall the commentator for the Cable News Network immediately began to talk about the sacred boundaries of the Soviets and how sensitive they are to Soviet violations. He was giving the Soviet defense before the Soviets themselves had made the

effort. The Soviets were silent for several days, and in the meantime all their excuses were publicized by Western journalists. Behind this is not an agreement with the Kremlin and a conspiracy. You have to give conspirators credit for having enough guts to conspire. We are talking about men and women in the news media who have no spines. The common thread that runs through the U.S. news media is fear, cowardice. They are afraid of the Soviets; they are afraid the people who are working for the Kremlin are going to win and we are going to lose. And when we lose, they don't want to be among those who are shot. They want to butter up to their new masters of the world.

This has happened before in history. I recall, for example, the incident of a newspaper in Paris at the time that Napoleon broke out of his first exile on the Island of Elba. The Paris newspaper's headline read: "The Monster Is Loose." And each day as Napoleon got closer to the French capital, the headlines became more modulated, and finally when he was on the outskirts of Paris the headlines read: "His Majesty the Emperor Will Arrive Tomorrow."[7]

Perhaps just as remarkable as the national news media's effort to find extenuating circumstances for the Soviets' behavior was their reluctance to believe that the Reagan administration was telling the truth about the circumstances that led to the destruction of KAL 007. For instance, the *New York Times* and the *Washington Post,* followed by the national networks' nightly news programs, implied that an RC-135 reconnaissance plane flying within seventy-five miles of KAL 007 might have given credence to the Kremlin's cover story that KAL 007 was a spy plane or that the Soviet military had mistaken the 747 for an RC-135. Almost no newspaper or national network took the initiative to have an experienced interceptor pilot sit down with the published radio intercepts and analyze them.

Retired veteran U.S. Air Force interceptor pilot Col. Samuel Dickens told this author:

I was both amazed and angry that no one in the news media was willing to do the basic job . . . required of such a story and ask an experienced interceptor pilot what the radio intercepts actually told us about the last moments of KAL 007 and whether in fact the Soviets did not know what they were attack-

> ing. After carefully reading the radio intercepts, there is no way in my mind that the radio intercepts coud lead to the conclusion that the Soviet fighters had mistaken the 747 for an RC-135.[8]

A consensus developed very early among members of the leading liberal news media not only that it was all a horrible Soviet blunder but also that moral outrage and demands for some kind of retaliation were dangerous.

Syndicated columnist Ellen Goodman wrote:

> A tragedy born of miscommunication, paranoia, and hair-trigger hostilities in a dangerous world is being used to increase the danger. If it were my mother, brother, child, friend, I also would be saddened to see these deaths escalate the possibilities of universal catastrophe. I would particularly be appalled to see the remains of the peace movement washed up on the political shores like grisly debris on the beaches of Japan.[9]

The pronounced appeasement-pacifist thinking revealed in the national news media coverage and editorial comment over the Korean airline massacre contained a hidden dimension that George Orwell first identified in March of 1947.

> Creeds like pacificism and anarchism, which seem on the surface to imply a complete renunciation of power, rather encourage this habit of mind. For if you have embraced a creed which appears to be free from the ordinary dirtiness of politics—a creed from which you yourself cannot expect to draw any material advantage—surely that proves you are in the right. And the more you are in the right, the more natural that everyone else should be bullied into thinking likewise.[10]

Since the 1960s and the onset of the Vietnam War, the U.S. news media have more and more become a partisan participant in an effort to shape and influence the decisions of the government and have played the part of a moral bully with the hunting license of the first amendment to the Constitution. The thousands of boat people of Vietnam and the mountain of skulls in Cambodia stand as a grisly record of repudiation of the peace movement and its news media allies. But has that ruthless record convinced them that they were profoundly mistaken in their judgment?

Columnist R. Emmett Tyrrell Jr., observed:

From the downing of the South Korean flight back to the fall of Vietnam, no movement in American history has been so soundly discredited by events. Will the consensus come to recognize it as an anti-defense-pro-appeasement movement rather than a peace movement? The only peace it offers is the peace of the grave or the cellblock.[11]

The U.S. national news media's consensus that Iran, Zimbabwe, and Nicaragua were in need of "change" led to the victory of totalitarian regimes far more ruthless and bloody than the hated pro-Western regimes they supplanted, which raises a profound and still unexamined question about the news media. It is a question that was raised by Ambassador Jeane Kirkpatrick in the aftermath of the KAL 007 massacre, and while the media and the Reagan administration were confronting one another over the future of the Caribbean and Central America.

The former academic-turned-diplomat pointed out that in El Salvador the U.S. news media routinely villify the government under siege while persisting in portraying the Marxist-Leninist terrorists as "social reformers." At the very same time, the bitter resistance to Soviet tyranny in Afghanistan suffers from calculated news media neglect. Ambassador Kirkpatrick observed:

The electronic media and mass education render culture manipulable in new ways. We are now touching on one of the most pressing questions of our time—one too shocking to raise. . . . Can democratic governments survive the systematic (and unsystematic) distortion of political reality by the press, radio, and television under conditions of mass media?

It is, above all, from culture that people derive their first and lasting conceptions of reality—their notions of what causes what, what is and what is not a worthy human endeavor, why some sort of experience is more worthwhile than another, why the monuments of human art and human learning deserve respect, and so on.[12]

Congressman Lawrence McDonald, long a critic of the national news media, particularly when it came to their reporting U.S. foreign policy, made the observation that the news media as an elite favor drawing all political power away from local, county, and state governments and placing it in the hands of the federal government—contrary to the intent and purpose of the framers of the Constitution. Helping in this centralizing process are liberals in Congress and in the federal judiciary.

McDonald told this author in June 1983:

> In the final analysis, the problem is in the living rooms of America. We have lost an informed electorate. Which leads me to the most powerful branch of government. The national news media now have the power to drum up interest in any issue and present it as "news." They, along with Congress and the judiciary, have contributed to using government as an instrument for social engineering that has led to a conflict-ridden society, as various groups compete for a piece of the federal pie. The news media play an important role in creating conflict and destroying the Constitution by editorializing in the news and creating the *appearance* that mass support exists for various social programs.[13]

We are confronted with a strange set of circumstances, which the evidence and experience of the last two decades clearly reveal. On one hand, we find a news media powerful and unaccountable, exerting great influence over foreign policy, even shaping decisions that have led to dictatorships—as in Nicaragua. On the other hand, we have a news media persistent in advocating a policy of appeasement of the Soviet Union in the name of sparing the earth from nuclear war—the assumption being that it will be the United States that will create the war! But in matters of domestic policy the political bias of the media is directed toward the promotion of greater centralization of power over the lives, liberty, and institutions of the nation, save for one institution—the news media itself!

On the eve of the year that George Orwell used to title his novel, 1984, several U.S. academics, intellectuals, and members of the news media observed that nothing like that which Orwell described in his nightmarish novel had in fact come to pass. This overlooks why Orwell wrote his work in the first place. What worried him the most was the lust for and love of power by intellectuals and a corresponding weakening on their part of a belief in liberty. He had observed that the centralization of power in fascism and communism—particularly in economic areas—was a warning for the future.

Shortly before his death in 1950 Orwell wrote:

> I do not believe that the kind of society I describe necessarily *will* arrive, but believe (allowing of course for the fact that the book is satire) that something resembling it *could* arrive. I believe also that totalitarian ideas have taken root in the minds of intellectuals everywhere, and I have tried to draw these ideas out to their logical consequences. The scene of the book is laid in Britain in order to emphasize that the English-speaking races are not innately better than anyone else and that totalitarianism, *if not fought against*, could triumph anywhere.[14]

23

The Widow Who Would Not Weep

"I just decided there was no way these people were going to see me cry."[1]

Mrs. Kathryn McDonald
November 15, 1983

The cold gathering gray mist in the Sea of Japan was the setting for a funeral farewell for six U.S. Navy and Coast Guard ships and two Japanese salvage vessels. On November 6, 1983, the eight search ships in single file turned away from Moneron Island off Soviet-occupied Sakhalin Island for the last time and set a course for northern Hokkaido and home port. After sixty-six intense days of searching for the wreckage of flight 007, the United States, Japan, and South Korea had decided that the sea would not give up the dead and the secrets of their final moments. The U.S. commander of the lead ship noted in his log that through the cold melancholy mist at least eight Soviet ships could be seen still searching off Moneron Island.[2]

Kathryn McDonald later told this author:

> All the debris that the Soviet Union boxed up and turned over to us contained, as far as I know, nothing of Larry's. But I want something physical, something concrete. The clerk of the House of Representatives wanted me to know that the House pays for a headstone and funeral expenses. But where can we put it? We don't have anything that proves to us that Larry is no more! It's as if he packed his bags, went into the elevator, blew me a kiss as the door closed, and got on that airplane and flew off. That's it! It's a problem that few have to face; it's death without the victim, a murder without a body.[3]

Forty-eight hours after the search in the Sea of Japan was officially ended, Kathryn McDonald struggled through the final hours of a grueling and at times cruel eight-week political campaign in Georgia's seventh congressional district. She ran in the special election to fill the vacant congressional seat of her murdered husband because on two previous occasions, according to Mrs. McDonald, her husband had asked her to make a run for the seat if anything ever happened to him.[4] The Georgia Democrat, according to his brother, Dr. Harold McDonald, had taken to wearing a bullet-proof vest on occasions after repeated threats had been made on his life.[5]

Winning the October 18 "first round run-off" in a field of eighteen candidates, Kathryn McDonald was decisively defeated in the November 8 run-off by losing four of the seven counties in the seventh congressional district to Georgia State Sen. George W. "Buddy" Darden.[6] Darden was the candidate of the Democratic regulars who had long disliked McDonald's views.[7]

Atlanta, which borders the seventh district of Georgia, prides itself on being compassionate; it rallied the entire nation when a procession of black children was brutally slain by a demented murderer. Since 1968 the city has treated Mrs. Martin Luther King, Jr., the widow of the murdered civil rights leader, with a respect and deference reserved for wives of slain heads of state. Despite the unpopularity of doing so, Congressman McDonald had opposed a national holiday for Dr. Martin Luther King, Jr.

The Georgia Democrat had also opposed the nomination of Andrew Young, now the mayor of Atlanta, as President Carter's ambassador to the United Nations, appearing as the sole opposing witness at his Senate confirmation. Congressman McDonald told the Senate Foreign Relations Committee hearing:

> If Andrew Young were not a principled man, he would subordinate his strong personal views to those of the United States and serve his country's interest at the United Nations. But Andrew Young is a man of principle and will of necessity find his own views coming in conflict with the United States' foreign policies. As he himself told the *New York Times* last month: "I'm going to be actively working within the State Department, the Congress, and the executive for my own concerns."[8]

From the time of his Senate confirmation in January 1977 to July 1978, Ambassador Young would prove McDonald's perception highly accurate. In early July 1978 Ambassador Young offered the provocative view that there were thousands of political prisoners in the United States. The then Senate majority leader Robert Byrd (D–W.Va.) told a nationwide television interview panel that if the U.N. envoy "makes another irresponsible statement that fits into this pattern of bad judgment, I think he should go."[9] On July 13, 1978, shortly after Ambassador Young's outburst, Congressman McDonald secured eighty-one signatures on a petition for Ambassador Young's impeachment, but it was defeated in a House vote.[10]

House Speaker Thomas "Tip" O'Neill (D–Mass.) told reporters later it would be best if "Andy curbed his tongue." Former Secretary of State Henry Kissinger found himself echoing Congressman McDonald in calling for Ambassador Young to resign if he could not keep his private "outrageous" opinions to himself. He said that Andrew Young should learn to "discipline" himself in his diplomatic duties or "not continue in his post."[11]

In August 1979 Mr. Young was finally forced to resign after initiating on his own talks with the Palestinian Liberation Organization, which were contrary to the Carter administration's policy. A *New York Times* editorial chose to conclude that while the U.N. ambassador's "candor" had both value and charm, his contacts with the PLO produced "a clumsy, foolish diplomacy that led his government into a lie, violated its policy, and broke its promises.

> Mr. Young's usefulness, great as it was, has been destroyed. . . . When a government speaks to issues of war and peace, it must have a reputation for speaking the truth. It can never tolerate the impression that it does not even *know* the truth about its own conduct.[12]

The Andrew Young affair was an illuminating illustration of what confronted Congressman McDonald for his entire public career in the Congress, where both he and Mr. Young had served prior to Mr. Young's appointment as U.N. ambassador.

The former aide to the late Dr. Martin Luther King, Jr., could commit what many finally concluded were the most outrageous offenses against common sense, reality, and the national interests; yet he continued to enjoy support, praise, and respect as a leader on the liberal or political left. But when Congressman McDonald was as candid and as forthright as Mr. Young, even courageous in expressing his convictions in the face of conventional wisdom, his voice and views were characterized as extreme, irresponsible, and even dangerous. However, in the final analysis the acid test for judging a public official, elected or appointed, is the reality of his or her record and what it says about the official's judgment—thus his or her competence to serve in office.

When Andrew Young's forced resignation as U.N. ambassador officially took effect, it was only a few weeks before Iranian revolutionaries seized the U.S. embassy in Teheran and held U.S. diplomats as hostages for fourteen months. It was an event that would bring down the entire Carter administration and help elect Ronald Reagan as president in 1980. Nine months before this shattering experience, the Ayatollah Khomeini was murdering supporters of the Shah in Iran. In February of 1979 White House Press Secretary Jody Powell was forced to chastise and disavow the statement by Ambassador Young: "Khomeini will be somewhat of a saint when we get over the panic of developments in Iran."[13]

When KAL 007 was destroyed in September 1983, Andrew Young had been elected mayor of Atlanta and former President Jimmy Carter was in Plains, Georgia, languishing in voter-imposed political exile. Both represented powerful elements of the Atlanta and Georgia Democratic establishment that Congressman McDonald had spent his entire public career successfully defying by getting himself reelected despite his prominence in the John Birch Society. The Georgia Democrat's repeated ability to overcome the powerful establishment in Atlanta both baffled and mystified the regular state Democratic organization and the Atlanta news media.

Tommy Toles, press secretary to Congressman McDonald and campaign manager in his widow's bid to complete the

Georgia Democrat's term, recalled the reaction of the Atlanta establishment and the Georgia Democratic party leaders when it was confirmed that Dr. McDonald had perished on KAL 007. Toles said they were "very, very restrained in their expression of regret" and could find very little to say about one of their native sons who had been so brutally murdered by the Soviets.

Toles noted, however, that when Mrs. McDonald made it official that she would seek her husband's congressional seat, the Atlanta establishment was actively involved in an effort to do to Lawrence McDonald in death what they had not been able to do to him in life. He added:

> All the rules of compassion and caring were suspended when Kathy McDonald announced she would seek her murdered husband's seat. For her to have been treated as viciously revealed the mean-spirited and petty nature of the people who think of themselves as compassionate, caring liberals.[14]

As an illustration, Mr. Toles pointed to a rumor that swept the district immediately after Kathryn McDonald entered the special election.

> The news media in Atlanta chose to ignore it, but a vicious rumor spread from one end of the district to the other like wildfire that Kathy had been trying to get a divorce from Larry. There were several versions of this lie, and it was something we were never able to shoot down in a district committed to family values. There was not much else with which they could defeat her, and Larry's enemies couldn't come out and attack her openly and honestly. Instead this vague rumor was used with devastating effectiveness as far as public perceptions were concerned.[15]

A major theme that Atlanta and the national news media did play on in the eight-week campaign was the refusal of Mrs. McDonald to play the role of the grieving, weeping Southern widow in public. "The TV people in particular," Mr. Toles said, "seemed to be very irritated that they couldn't get pictures of Kathy crying. It was just amazing. It seemed as if they were trying to spread the false impression that she didn't care."[16]

A survey taken shortly after the October 18, 1983, first round in the special election in the seventh district of Georgia indicated that Mrs. McDonald was perceived by 60 percent as unqualified and as "an opportunist," because she was running on the memory of her murdered husband and not in her own right.[17] The news media's emphasis that she was from California, although she had lived in the district since her marriage to Dr. McDonald in 1976, combined with her refusal to display her grief in public all added up to Mrs. McDonald's being perceived as a cold, calculating female. The news media came to dub her "the Ice Lady," to which Mrs. McDonald replied with the directness and candor characteristic of her husband. "I have been called 'the Ice Lady,'" she told a group of reporters in Georgia shortly after announcing her candidacy. "There's no accounting for the bad taste some people will exhibit."[18]

Kathryn McDonald told this author after her defeat that like her husband she made little effort to disguise her contempt for most of the local and national news media. During his nine years in Congress most, but not all, reporters exhibited little real, intelligent interest in what McDonald was doing and why he took the positions he did and whether in the end he was proven right or wrong.

Mrs. McDonald told this author:

> They were absolute animals after Larry was murdered on KAL 007. They camped on our doorstep in Georgia at all hours of the night, ringing the security buzzer and waking up our two small children and whining to have me come down to make a statement so they could meet their deadlines. And when I did oblige them, they stuck microphones and cameras in the face of my two-year-old, Larry, Jr., and asked him if he knew where his father was.
>
> I have often noticed in the coverage of tragedies by the news media that reporters and TV cameras focus in on the grieving person, to squeeze some emotional reaction out of them. Look at what they did with those poor parents and wives of the U.S. Marines murdered in Lebanon. Before those poor people knew the fate of their loved ones, the TV people and print media were playing for the emotional; and they couldn't wait to get some kind of emotional reaction to sell newspapers and hype the TV

news show rating. I just decided *there was no way* those people were going to see me cry. I conducted myself the way Larry would have wanted me to. I am not sorry I did, although I know it hurt me in the campaign. I have always been a private person, and my feelings are mine and not the business of strangers. I could never hold public office for any length of time and I could hardly stand campaigning, particularly the press prying into my personal life. I despise them for it.[19]

Georgia State Senator Darden had everything going for him in his successful race against Kathryn McDonald: backing of the Atlanta news media, organized labor that poured money and manpower into his campaign, the state Democratic organization and, perhaps more important, nine long years of resentment and hatred of McDonald by his own party because the congressman chose to stand alone. Coursing through the underground of the Darden-McDonald race, too, was the resentment of the Carter-Young faction which could not forgive Larry McDonald for his relentless opposition to their policies while they were in power in Washington and, in the end, for being proven right that their policies were disastrous for the country and their party.

Kathryn McDonald observed:

The sad thing is that they think they are the intelligent, sophisticated set. Many of the people who were against Larry should know better. They didn't vote for me because I was not a professional politician. If only people would realize, as Larry stressed over and over again, we already have too many professional politicians who want and need the job and will do anything to get it. We need people to run for public office who don't need the job. The American republic was founded by men who didn't need or want to become professional politicians; they had the job thrust on them because they believed they had a duty and obligation to their country. What we have today are 245 lawyers in Congress. We are becoming a nation of politicians guided by politicians for politicians. They are very adept at getting elected by being all things to all people and end up with no principles and no self-respect for themselves or the people who elect them.

Larry McDonald was different because he respected himself, he respected the people who sent him to Congress, and he knew how and why this country became great. People used to be puzzled why Larry seemed always to be voting against the majority in Congress. Larry always judged each issue by three basic guidelines: First, is it Constitutional? Second, can we afford it? Third, do we really need it?[20]

24

A Don Quixote from Dixie

***"We are producing a conflict-ridden society by the same process that destroyed past nations and civilizations."*[1]**

U.S. Rep. Lawrence McDonald
June 2, 1982

"Southerners have been the architects of the American political system," observed historian David Leon Chandler, "and during two centuries have functioned as its caretakers, working to keep the system close to the original design."[2]

The original constitutional design in the document of 1789 was that limits should be placed on elected representatives' use of power over the life, liberty, and property of the individual and that the system should be a government of laws, not of men. As a native Southerner born in Georgia, Congressman Lawrence McDonald was a product of the Southern political tradition. Originally it had been the national political tradition; the Declaration of Independence was written by Thomas Jefferson and the Constitution by James Madison, both Southerners. It is now largely ignored or forgotten that, while the North gave America its capacity for economic greatness, it was Southern leaders who created the democratic structure that made the United States great politically.

Two centuries later, as Chandler points out, the worst fears of Thomas Jefferson and James Madison have been realized in the growth of government power and of an entrenched bureaucracy unanswerable to elected representatives. Southerners for the last century have refused to agree that the growth of government was inevitable. "They have resisted every step of the way," notes the Pulitzer Prize-winning writer.[3]

Fourteen months before his murder, Congressman McDonald told this author that the republic created in 1789 is today a political fiction, its genius gutted by Congress, the White House, the courts and the media.

> It has in fact been the problem for fifty years or longer. Sixty percent of the Federal budget is devoted today to redistribution of income. Now the Founding Fathers and the framers of the U.S. Constitution worked to prevent government from becoming what it is today. So in the most realistic sense, the U.S. Constitution as far as Congress and the courts are concerned is nothing but a dead-letter document.[4]

While Congressman McDonald was willing to face the reality of this largely evaded historical tragedy, it did not prevent him from advocating throughout his congressional career a restoration of the American republic. He wrote in his 1976 book:

> Pleading for reestablishment of constitutional government, as ordained by the Constitution of 1789, is not ignorant advocation of a return to "a simpler world." Rather, it is a suggestion that the nation climb up and stand on the high ground it once proudly held. The high ground is still there. The governmental system ordained in our Constitution is as appropriate for our nation today as it was for America of 1789. It is even more desperately needed now than then.[5]

Congressman McDonald was concerned about the emergence of a new slavery in the United States, the growth of the all-powerful state at home and the refusal of the Christian West to wage a concerted moral, intellectual, and theological war against the slave-state mentality of atheistic Marxism-Leninism abroad. He saw America's salvation as coming from the restoration of the republic and the religious heritage from which it originally drew its moral and intellectual strength.

Dr. Daniel Jordan, a friend and fellow physician for twenty-five years, maintains that originally Dr. McDonald did not throw himself into the struggle that consumed his life out of an ideological commitment. Dr. Jordan told this author:

> It did not begin, as many assume, out of ideological concerns, but out of his experience as a trained medical physician. When

> he returned from Iceland in 1961 he had seen firsthand, as the Naval physician for the U.S. Embassy, how the diplomats there were responding in all the wrong-headed ways to the apparent Communist influence in Iceland. It was also while he was stationed at the Naval hospital at Bethesda, Maryland, that he saw firsthand what the political system did physically to U.S. congressmen and senators. It was a combination of both of these experiences that triggered his decision to become involved in politics. He became acutely aware of the problems the country was going to face in the future and resolved to do something about it.[6]

Dr. McDonald looked on American society and later on Western civilization much as a physician would look on a patient suffering from a potentially fatal disease. What was required first was an extensive examination and diagnosis of the body politic. The history of the United States and the West was to him much like the medical histories of patients whom he, his father, and his brother treated in their joint medical practice.

He joined the John Birch Society, according to Dr. Jordan, after discussing his experiences in Iceland with a former Georgia congressman who urged him to look into the group (it was generally regarded as extremist and even dangerous in the early 1960s). His membership in the society, which led to the breakup of his second marriage to an Icelandic national who did not share his intense views of the crisis facing the West, was to demonstrate very early his indifference to—if not a disdain for—what others might think. Dr. Jordan explained:

> He took the political route because he lacked any real alternative that had some prospects of achieving results. After a while he realized that he was unique and had an amazing ability in the political arena. No one would challenge the incumbent in the seventh district of Georgia. [McDonald] did and lost in 1972 but surprised just about everyone by winning in 1974. He never exhibited a sense of defeat and hopelessness that had been characterized in the conservative movement. He was always victory- and action-oriented. In one sense he thrived on controversy. He was never a worrier and was more a formulator, constantly trying to work to get something accomplished in the long range.[7]

During the nine years he sat in the House of Representatives, Larry McDonald used his seat as a political pulpit for his long-range goal of bringing unity to a divided and demoralized conservative movement. He built among non-Birch Society conservatives a national reputation for political honesty and integrity and a willingness to work with others for the common goal of overthrowing the dominant liberal establishment by the ballot—not the bullet, as so many of his enemies believed he secretly advocated. "Others debate public policy," observed *Atlanta Journal* columnist Dick Williams after his murder. "He revered the Constitution. He was a literalist, a fundamentalist. That simplicity, often dismissed as buffoonery, endeared him to the religious fundamentalists of the seventh district."[8]

The Rev. Joseph C. Morecraft III, minister and friend of Dr. McDonald, suggested that his appeal was much deeper and more significant, exhibiting by his lonely independence and magnetic personality something of the qualities of the founders of the republic. Morecraft observed:

> The interesting thing about north Georgia is it is predominantly rural, mountainous—the southern tip of the Appalachian Mountains, with Scot-Irish ancestry. It still maintains a Christian consensus, and these people still have a memory of the kind of life and values that they learned from their grandfathers and fathers. When he was murdered, people streamed into the district office—grown, older men with tears in their eyes—as if they had lost a kin. Not all of them could verbalize their feelings, but what they and we all knew was we had lost a man who knew something was wrong and tried in his own way to make it right. He reminded them of the wise ways of their grandfathers.[9]

Even his most bitter enemies in Congress and the media granted that he did not fit the stereotyped Birch fanatic that had been fabricated ever since the society was founded in the late 1950s. Joseph Nocera, associate editor of the liberal magazine, *Texas Monthly,* wrote after Congressman McDonald's murder that it was not hard to understand why he was repeatedly reelected to Congress. Besides being a first-class campaigner, he was a polished, made-for-TV politician. Mr. Nocera wrote:

> He was unbeatable in campaign debates; always calm and collected, always sounding completely rational as he talked about the issues upon which he and his constituents agreed—the need to balance the budget, say, or to increase defense spending. Invariably, it was his opponents who became unhinged as they tried, unsuccessfully, to tar him with the Birch brush.
>
> If there is anything admirable about McDonald's five terms in Congress, it is that he practiced what he preached. Unlike many of his more "spineless" colleagues (McDonald's favorite adjective when discussing Congress), he never made exceptions for his own district. Because he opposed all nondefense federal spending, he consistently voted against bills that would have pumped money into his own district.[10]

Fellow Congressman Newt Gingrich (R–Ga.) concluded that while he was effective nationwide, he was ineffective in Congress. However, former House staffer and now strategic scholar Bruce Herbert believes, after watching Congressman McDonald over the years, he was effective in raising critical issues that others were too timid or afraid to approach. Herbert told this author:

> He would raise an issue or introduce legislation which nobody supported until the vote came. He was an excellent sparkplug. I think this recognition and credit was withheld because many did not want to be labeled a right-wing kook. But when it came time to vote, on many occasions they went along with McDonald. He had a great deal more influence than he is generally credited for having achieved.[11]

Kathryn McDonald revealed that both she and her husband's Congressional staff saw the Georgia Democrat as a Don Quixote. The character in Miguel de Cervantes's epic novel was fired with notions of romantic chivalry from books. He believed them and went forth to tilt at windmills, which he thought were giants.

"He admired all those who dared to be different, who fought the odds. We used to laughingly refer to Larry as Don Quixote, a few of us who knew him well," Mrs. McDonald told this author with a smile. "I felt like he was fighting windmills."[12] He had in his Washington office stuffed dolls of Don Quixote and Sancho Panza, exhibiting a wry sense of humor seldom guessed at by the press.

However, unlike the Man of La Mancha, Congressman McDonald's analysis of the crisis confronting Congress, the nation, and the West was more than mistaking windmills for giants. In the 1980s, he told this author, the erosion and disappearance of the constitutional checks on the exercise of political power by elected and appointed representatives had led to the conflict we are witnessing in our domestic society.

> As limits upon government have broken down, as these limits have been destroyed, the federal government in essence may do anything that Congress says it thinks it ought to do. The federal government today, thanks to both Congress and the White House for most of this century, is involved in almost every sphere of human endeavor. Not only in this country but around the world. As a result, we have conflict between various people who want people's taxes to be used for this project or that project. In the final analysis, these competing groups, which politicians in Congress cultivate, must clash and create conflict.
>
> We are producing a conflict-ridden society by the same process that destroyed past nations and civilizations. We have a fragmentation process going on because we have allowed the slow destruction of the U. S. Constitution. And the so-called House leadership of my party has been in the vanguard of the destruction as a limiting document. It is not a very nice or noble present for the American people for the two hundredth birthday of the U.S. Constitution [in 1987]![13]

In the aftermath of the KAL 007 incident, few if any of the news media ever referred to the Georgia Democrat as an idealist who had hopes, dreams, and aspirations for his country and the Western world. This denial, deliberate or out of plain ignorance, is contrasted to the idealism accorded to murdered political figures such as President John F. Kennedy, Sen. Robert Kennedy, and Dr. Martin Luther King, Jr.

Paul Weyrich, director of the Committee for the Survival of a Free Congress, a conservative grass-roots organization on whose board Congressman McDonald served, noted that if

Dr. King and his followers had a dream, why was it so hard to believe that the Georgia Democrat also had a dream?

Four days after the destruction of KAL 007 Mr. Weyrich wrote:

> Larry McDonald must not have died in vain. The innocent citizens of America and other countries who were slain must not be forgotten. Those of us in the conservative leadership have resolved to make Larry our benchmark for integrity and clarity of purpose. Conservatives around the country should undertake activities to sustain the vision of this good man. This is one Soviet atrocity we must not allow to fade from popular consciousness.
>
> It was typical of Larry that, when these murders occurred, he was on his way to an international conference celebrating the continued freedom of Korea, thirty years after the signing of the armistice there. School children across America should learn about these obscene Soviet murders . . . and about Larry's vision—just as they are taught about other crucial moments in American history.
>
> It is an irony that this patriot and friend of liberty died on the two hundredth anniversary of the signing of the Treaty of Paris, the document formally ending the War for American Independence. Larry McDonald lived and died to sustain the ideals which gave birth to that glorious revolution.[14]

Epilogue

The Twilight of the West? An Essay on the Evidence

***"If evil, villany or war comes, and if the Watchman does not sound the trumpet, then the blood is on his hands. . . ."*[1]**

Dr. Lawrence P. McDonald
Marietta, Ga., Speech
June 1983

Beginning in 1987 the United States will mark the two hundredth birthday of the Constitution. If the bicentennial celebration of the Declaration of Independence in 1976 was any guideline, efforts will be made to pretend that in practice and in principle the U.S. Constitution is "a living document," rather than one that Congressman Lawrence P. McDonald termed "a dead letter document."[2]

The crime against the Constitution has been committed over several generations, gutting its original genius of limiting the exercise of political power over the liberties, lives, and property of individual citizens. It has been committed in the belief that human problems of poverty, discrimination, and other societal ills could be solved by the exercise of political power by an elected and appointed elite using public tax monies.

The contemporary and historical evidence clearly repudiates that such a philosophy of power, as contrasted to the philosophy of liberty of the founders of the American republic, has effectively served both the commonwealth and the commonweal. Instead, conflict, disorder, and corruption have been the consequences of this policy. The founders of the American republic saw clearly how ancient civilizations were destroyed by such a philosophy, and based on historical evidence, they devised a constitutional system to prevent its repetition.

However, both the contemporary and historical evidence that this philosophy of power is systematically destroying domestic concerns is largely ignored or evaded by both political parties, academics, intellectuals, businessmen, labor leaders, and the news media. The reason for this bigotry toward evidence and experience stems from several sources, the first being the subjective nature of political leaders and the system itself, which is subjective and irrational rather than objective and rational. The second cause for the evasion of evidence and experience is intellectual laziness and cowardice in facing the facts, out of the fear they may contradict some cherished notions about life and human nature. Underlying most if not all the proposals for the expansion of powers of government over all areas of human endeavor is the belief in the idea of equality. The founders of America had concluded that if there was one doctrine that would spell the ruin of the constitutional system they created, it would be the belief in a mass democracy founded on the idea of equality of condition as opposed to equality of opportunity.

James Madison, writing in the Federalist Papers after the creation of the Constitution, left no doubt as to his views and those of the other Founding Fathers when it came to the ideas of mass democracy and equality.

Madison wrote in Federalist 10:

> Democracies have ever been spectacles of turbulence and contention; have ever found to be incompatible with personal security or the rights of property; and have in general been as short in their lives as they have been violent in their deaths. Theoretical politicians, who have patronized this species of government, have erroneously supposed that by reducing mankind to a perfect equality in their political rights, they would at the same time be perfectly equalized and assimilated in their possessions, their opinions, and their passions.[3]

The historians Will and Ariel Durant after concluding their epic life-long study and writing of the multivolume work, *The*

Story of Civilization, summarized their conclusions in a slim volume, *The Lessons of History.*

> Inequality is not only natural and inborn, it grows with the complexity of civilization. Hereditary inequalities breed social and artificial inequalities; every invention or discovery is seized by the exceptional individual, and makes the strong stronger, the weak relatively weaker, than before. Economic development specializes functions, differentiates abilities, and makes men unequally valuable to their group. . . .
>
> Utopias of equality are biologically doomed, and the best that the amiable philosopher can hope for is an approximate equality of legal justice and educational opportunity. A society in which all potential abilities are allowed to develop and function will have a survival advantage in the competition of groups. This competition becomes more severe as the destruction of distance intensifies the confrontation of states.[4]

The liberal Welfare State created in the United States in the last half century is rooted in the idea of equality of condition. While this has been the *intent*, its actual consequences, objectively, have been to create group conflicts, to weaken the liberties and opportunities of all citizens, and to sow the seeds of discontent and disbelief by all citizens in an entire spectrum of institutions—government, education, business, labor, and communications. We are as a nation weaker internally in the 1980s than we were twenty-five years ago. Conflicts, distrust of leadership elites, and a general lack of national unity are more manifest as we approach the two hundredth birthday of the U.S. Constitution than only a generation ago.

This internal weakness caused by conflicting groups competing for the spoils offered by the Welfare State bears directly on our ability to survive domestically with our freedoms unimpaired, while internationally we are faced with an enemy who believes in power, not law, nor liberty, and certainly not in the preciousness of life itself.

This brute reality was vividly and tragically illuminated by the Soviets' destruction of Korean Airlines Flight 007. What were also illuminated, which this work has sought to demonstrate by detailed documentation, were the weakness, fear, and even cowardice exhibited by the leaders of the United States. They firmly refused to deal realistically with the barbaric behavior of a system that means to destroy what their country has stood for ever since its first pioneer settlements in 1607.

With such leadership, our capacity for survival is clearly in question as we approach our two hundredth year as a free people who began with a system of laws and not men, and which has been transformed into a nation ruled by men, not laws. Congressman McDonald had put forth an analysis of the crisis confronting us and had proposed certain steps of intellectual, philosophical, and theological restoration to meet the crisis that is progressively overcoming Western civilization. During his nine years in Congress his vision for a revitalized America was either not understood or deemed unrealistic for a modern era. The majority of his colleagues seemed to feel that ideas and concepts that took root in the age of gunpowder and horses are hopelessly unrealistic for a time of electricity, computers, and nuclear weapons.

However, this amounts to an evasion rather than an answer. In order to gain the right answers about human problems, one needs to ask the right questions. The paramount question of our time is whether, if in viewing all of the evidence and experience, we can really believe that the current course on which the United States has been proceeding in the last generation will guarantee our survival at home and abroad. Congressman McDonald thought not; and considering the accuracy of his predictions on domestic and foreign policy issues, he may very well have been right in his belief that without fundamental changes the United States as a free nation is doomed.

Congressman Newt Gingrich (R–Ga.), before his election to the House of Representatives, taught history and received his doctorate in the subject. While he does not share all the conclusions of his late friend and colleague, he has fundamentally come to the same conclusion that the United States' survival capacity is highly doubtful in the long range. Gingrich told this author:

> Without adequate national leadership we face a real problem in that the liberal Welfare State is designed for impotence in international relations; its model of how to deal with international problems is entirely wrong. The liberal Welfare State does to power what the Victorian era did to sex—it hides it and denies it. As a consequence, it is very hard in this society to talk rationally and calmly about the nature of the world and the nature of evil. Candidly, unless we have a fundamental cultural change, I think we are finished. I think the survival probabilities, so long as the United States remains a liberal Welfare State, are under fifty years.[5]

The enormous public expenditures of the Welfare State have often been at the expense of a strong national defense. The advocates of disarmament either deny, or offer a wide range of excuses for, the Soviets' murderous conduct. This misperceived view of political criminal conduct is rooted in the same delusion that dominates misconceptions as to why domestic criminals commit crimes against the law-abiding.

In his forthcoming book, *On Survival,* Congressman Gingrich argues that

> the values, habits, institutions and framework of thought dominating our society are wrong about the world at large. They underestimate the dangers to our nation. They have consistently been wrong about Soviet and Third World behavior for the last twenty years.
>
> Furthermore, the dominant cultural values have proven wrong about human nature in domestic, economic and social policies. The centralized bureaucratic model of government has weakened our school system, our criminal-justice program and our economic health. Many defense procurement problems have worsened precisely as centralized bureaucracies force more regulations and red tape in response to problems. Solving one set of problems has led to even more difficult problems as we continue to apply the wrong rules of behavior, treating symptoms while the disease is largely cultural.[6]

An illustration is contained in the presentation by the ABC television network of "The Day After," a fictional account of what it would be like to survive a nuclear attack. More public discussion, political concern, and news media attention were given to this fictional presentation of what *might* happen than was accorded to an event that *did* happen—the brutal mass murder of 269 innocent passengers aboard an unarmed airliner less than ninety days before the nationwide screening of "The Day After."

What this graphically illustrates is the capacity of a cultural mechanism like television and the news media to distort reality. For the brute reality of our time is not the *possibility* of nuclear war destroying all life on the planet, but the *actuality* of the use of terrorism and violence to achieve political objectives by the Soviets and their surrogates. Consistently since the early 1960s the U.S. and international news media have misrepresented the nature of this clear and present danger, devoting little if any serious analysis to its long-range implications. Furthermore, the news media have more and more given time and space to the views of those who have been the architects of previous domestic and international disasters.

In the wake of the destruction of KAL 007, for example, former Secretary of State Henry Kissinger's opinion was sought as to why the Soviets destroyed KAL 007. It stemmed, he said, from a failure of Soviet air defenses. "The accident itself was brutal," he said, "barbarous, inhuman, whatever you want to say, but I did not think it was an act of high policy."[7]

The record of accuracy of Mr. Kissinger's views and policies has never been the subject of objective and extensive scrutiny by the news media. For reasons that remain a mystery, the part that former President Richard Nixon and Dr. Kissinger played in the Vietnam "peace" of 1973 that led to the tragedies of the Vietnamese boat people and the genocide in Cambodia has been consistently ignored. Instead, the news media have accepted at face value Dr. Kissinger's contention that the Vietnam "peace" fell apart because of Watergate. But underlying that "peace" and hidden from public view has been Dr. Kissinger's own worldview that is shared by far too many U.S. political, intellectual, and moral leaders.

Ironically, during the bicentennial of the Declaration of Independence in 1976, only a year after the U.S. withdrawal from Vietnam which left its people to be victimized by a ruthless dictatorship, former Chief of Naval Operations Elmo R. Zumwalt, Jr., published his memoirs. Admiral Zumwalt maintains that President Nixon, through his secretary of state, sought "foreign policy successes" as a means to distract public attention from Watergate. With great intellectual and moral courage, he declined to be reappointed to the top flag officer position and took early retirement from a career he loved in order to write what he knew about the worldview of leaders like Nixon and Kissinger.

Admiral Zumwalt wrote:

> What is important to record is the inextricable relationship the Nixon Administration's perversion of the policy-making process bore to its ignoble outlook. Its contempt for the patriotism and intelligence of the American people, for the Constitutional authority of Congress, and for the judgment of its own officials and experts reflected in Henry Kissinger's world view: that the dynamics of history are on the side of the Soviet Union; that before long the USSR will be the only superpower on earth and the United States will be an also-ran; that a principal reason this will happen is that Americans have neither the stamina nor the will to do the hard things they would have to do to prevent it from happening; that the duty of the policymakers, therefore, is at all costs to conceal from the people their probable fate and proceed as cleverly and rapidly as may be to make the best possible deal with the Sovet Union while there is still time to make any deal. The political-military policies of the Nixon Administration flowed quite logically from that view of the world, which the President certainly went along with whether or not he had arrived at it independently of Kissinger. I think what I learned during four years in the thick of that miasma made it my duty to write a book.[8]

In the last eight years since Admiral Zumwalt made this startling revelation (one never pursued by the media), based on firsthand conversations with Kissinger, this view has undergirded U.S. policies of appeasement and was clearly reflected in the Reagan administration's handling of the KAL 007 incident. Such a view was also reflected in the refusal to take immediate retaliatory measures against the truck-bomb murder of over 230 Marines in Beirut, Lebanon, in October 1983. Such a policy of inaction in the face of a criminal terrorist act was ratified by former presidents Ford and Carter when, on November 6, 1983, they each cautioned that the United States should avoid military retaliation for the bombing of the Marine headquarters![9]

The persistent refusal of the government to use measured military means to safeguard vital and strategic U.S. interests and even human lives, out of a fear that such measures will eventually lead to a nuclear conflict with the Soviets, undercuts the demand that the United States rebuild its national defenses. Billions are earmarked for nonnuclear defense systems that the leadership is afraid to use in defense! Those advocates of the Welfare State, besides urging disarmament, argue that such resources could be better employed in domestic programs at home. The programs they advocate draw more power to the center and dictate the lives of millions while creating ever greater cycles of conflict among competing groups.

The consequences of current totalitarian trends in U.S. domestic life combined with the appeasement of a ruthless totalitarian world power abroad is not difficult to discern. Dr. Henry Kissinger, for example, was appointed by President Reagan as chairman of a Central American Task Force. "Henry Kissinger has no permanent place in the Administration," President Reagan insisted in an October 3, 1983, letter to former New Hampshire Governor Meldrim Thomson, Jr. "He is chairing a bipartisan study group which meets a Congressional demand as a price for getting the support we must have for our effort in Central America."[10]

Again, such an effort recoils from direct military involvement and, instead, advocates a negotiated settlement of conflicts whose origins were very early and accurately foreseen by Congressman Lawrence McDonald and others. Those warnings were ignored, and actions to insure the survival of allies in Central America were delayed until it was too late;

and the Cubans, Soviets, and their Eastern European and Asian surrogates gained a strategic position in Central American countries like Nicaragua.

In one of the last published articles he wrote on Central America, Congressman McDonald foresaw the domino effect, dismissed in Southeast Asia until it happened (and then it was ignored), coming into play in Central America and this hemisphere. He wrote in June 1983:

> The fate of El Salvador is in the hands of Washington. If we feed El Salvador to the Communists, they will simply move on to Honduras and Guatemala, and after that to Mexico. Which would be a disastrous finale to a process that began when American "Liberals" decided that Fidel Castro was just an "agrarian reformer."
>
> Our first and, overall, largest concern, has been with the deterioration of Mexico, with its increasingly corrupt, hostile, socialist Government. . . . Our own media are amazingly reluctant to tell us what is happening in that country with which we share a boundary that we would rather not see converted into a Cactus Curtain. There is little question but that Mexico is the final prize in the high-stakes game now under way.[11]

Beyond the flood of millions of refugees who would seek admission into the United States from a Marxist Central America and Mexico, such a development would place the United States in the position of being itself the eventual victim of a terrorist campaign conducted from Mexico that could progressively cripple the vast agricultural potential in the Southwest and the Southern California region. Clearly 50 percent of all our nation's produce would be vulnerable to terrorism from without and from growing internal disorders fanned by domestic radicals who desire to transform our system of government into a more radical and state-controlled one in the alleged name of equality.

The historian Otto J. Scott believes that the United States suffers from all the fatal symptoms that afflicted and overwhelmed the Roman Empire, the Spanish Empire after the sixteenth century, and the British Empire in the 1930s. Fear and an absence of strong religious values, he argues, are at the root of America's retreat from greatness, power, and responsibility since the end of World War II.

Mr. Scott observed:

> As we lose our allies, this will contribute to our paralysis of will until we face the same problem of terrorism at home that we have chosen to ignore in other nations or explain away as the consequences of poverty and social inequality. Finally, the dog whistle will blow and then we will have terror here at home. Until now they have held off knee-capping, murder, kidnapping as in Belfast and Italy. When we are finally surrounded and have no allies left, and Mexico is gone, then the terror will start; they have the instrumentation, they have the tools that other countries don't have, because we have minority races who have been encouraged into attitudes of resentment. Not, of course, the majority of them: the bulk of American blacks, Hispanics and other minorities in the United States are law abiding, decent and don't want any more trouble than anyone else. But the firebrands, the radicals and revolutionaries among them, the ones who have been selected and trained for leadership, the American version of the Ho Chi Minhs, will rise and we will have our Viet Cong, we will have *our* terrorists; we will have our period of destabilization. And they will call it, what? A civil war![12]

During the past two decades the attitude toward domestic criminals has been one that greatly jeopardizes the safety and security of the law-abiding. What would be the attitude of the U.S. leadership toward the domestic political criminal who wages terrorism? During his congressional career Larry McDonald argued for reconstituting a U.S. Internal Security system to meet the possible eventuality that the United States, like other nations, would be the victim of concerted terrorism. His critics assailed him for such proposals as advocating a police state over civil liberties.

As early as July 1978 Congressman McDonald was pointing out that almost all local, state, and federal internal security intelligence programs had been dismantled and that the Federal Bureau of Investigation and the Central Intelligence Agency had been portrayed as menaces to American freedom. He pointed out, by way of contrast:

> Since 1976 virtually every Soviet- and Cuban-backed Marxist terrorist group in Latin America, Africa, and the Middle East has set up, with the help of the Communists, often via agents assigned to the United Nations missions in New York City, support networks of academics, street activists, radical lawyers and journalists, and others. These support networks form a platform for possible future terrorist attacks by foreign terrorists. But it is also possible that certain individuals and small groups of supporters may perform acts of terrorism on behalf of their foreign terrorist allies since U.S. supporters can move about without attracting notice and are free of surveillance by intelligence agencies. The violence and terror are only a matter of time unless we act now to restore the House Committee on Internal Security.[13]

Until McDonald's murder five years later, most of his congressional colleagues, and the media, thought him paranoid and refused to authorize the creation of the House Committee on Internal Security. Besides, the fear of a nuclear conflict regularly fanned in the Congress and the news media has tended to paralyze clear thinking on the reality of the existing problem, while the focus of fear is on a problem that might happen.

Former top CIA official Raymond Kline put the problem succinctly in the aftermath of the film "The Day After" when he observed: "We are talking about the danger of long-range missiles in the 1980s when the one weapon that can take countless lives is a truck loaded with dynamite" driven by a terrorist bent on a suicidal mission. "We haven't kept our intelligence records on terrorists in this country for five years because we believe that it won't happen here as it has in other countries. But we are vulnerable to the same kind of campaign."[14]

Shortly after Mr. Kline's statement, the State Department and White House placed dump trucks loaded with sand at their main entrances after intelligence reports indicated that terrorists would try to do in the States what they had done to the Marines in Beirut! Additionally, a terrorist bomb had gone off in the U.S. Capitol building, which produced talk about a fence to close off the entire Capitol grounds!

Furthermore, President Reagan was forced to deliver his January 1984 State of the Union address to Congress under security measures reminiscent of war time. "Not since World War II," wrote Don Phillips of United Press International, "when machine guns were mounted on the roof and soldiers patrolled the steps, has the Capitol been as well protected as it was for President Reagan's State of the Union."[15]

In his State of the Union address there was no direct reference to the global problem of terrorism, to Central America, or to the destruction of KAL 007. "The United States is safer and stronger," the president insisted," and more secure in 1984 than before. We can now move with confidence to seize the opportunities for peace—and we will."[16]

Such a statement is challenged by how the president, the Congress, the news media, and the leadership elite in the United States chose to conduct themselves after the destruction of KAL 007. An illuminating incident, it brought into sharp focus for a brief moment the existence of a great and grave leadership weakness. The destruction of KAL 007 was a clear act of terrorism, despite the persistent denials by the leadership elite in America and the West that it was an accident, a Soviet blunder.

Lawrence Patton McDonald's description to this author in March 1983 that Yuri Andropov was like a deadly cobra who mesmerizes his victims before striking was not some idle graphic metaphor. What Dr. McDonald was describing was a process of fear that has taken root in the minds of Western leaders who are unwilling to face the consequences produced by a prolonged paralysis of will in the face of danger.

The *Day of the Cobra* is a shorthand description of a man, a system, and a set of ideas that Western political, intellectual, and moral leaders face dangerously disarmed. They face such a system lacking the proper intellectual and theological armaments to defend our system against an evil that has dedicated its very existence to our destruction, and which has not demonstrated in over a half century any deviation from that objective.

The *Day of the Cobra,* beyond the destruction of an unusually brave human being and 268 other innocent lives, remains a forecast of the possible Twilight of the West. KAL 007 demonstrated far more fundamental truths about the Christian Free West than it did about the atheistic East. It also illuminated a problem that we will have to face if we expect to survive as a free nation and as a civilization. Namely, that the leaders in the West who hold in their hands our lives and liberties have left us naked and defenseless before our enemies because they have lost faith in themselves and in all of which in the past had been a source of strength.

One of those main sources of strength is religious faith. The events before, during, and after the destruction of KAL 007 illustrated the fundamental weakness of the West: a fanatic faith that life at *any* cost is the highest value, because our leaders cannot bring themselves to believe that there is another life beyond this world.

Dr. Lawrence Patton McDonald in the last few years of his adult life had, because of his medical training, come to the conclusion that Western civilization in general and the United States in particular was very much like a gravely ill patient. His diagnosis was based, not on ideological grounds as his critics maintained, but on a body of historical evidence and experience reaching into antiquity.

A review of the *Congressional Record,* his speech files, correspondence, and his published articles in the short nine years he was in the U.S. Congress provides a body of evidence to judge just how accurate he was on a wide range of domestic and foreign issues. His analyses, as we have shown throughout this work, indicate an accuracy that few who sat in the same Congress can equal. If he was accurate in his analyses of contemporary, short-range issues, the question remains, How much credence can we place in his long-range prognostications? He had concluded that our survival as a free nation and civilization was in serious jeopardy unless we were willing to return to those principles that first gave birth to the Christian West and the United States.

Dr. McDonald's world view was woven out of the fabric of

evidence and experience that embraced not only political economic freedom but also theological convictions. First and foremost he shared with the founders of the United States the conviction that a nation, in addition to economic and political principles, needed theological foundations.

Congressman McDonald's ancestors had immigrated to America from England and Scotland prior to the American Revolution. As historian Otto J. Scott points out, the theological basis of the American Revolution was the deeply held conviction by the American colonial followers of John Calvin (1509–1564) and John Knox (1514–1572) that the British government in the 1770s was un-Christian because it had become unlimited. The state had assumed the powers of God and it was, therefore, the duty of every Christian to oppose such a rule.

In history Dr. McDonald admired such authentic American hero warriors as Gen. Thomas "Stonewall" Jackson, Gen. Robert E. Lee, and his distant cousin Gen. George S. Patton, Sr., of World War II fame. All three were from Virginia and were imbued with a Christian faith. Although born in Georgia and the son and grandson of physicians, Dr. McDonald was drawn to the calling of the warrior.

In the brief nine years he was in Congress it was a different, difficult, and more dangerous kind of warfare he chose, fought primarily with brains rather than bullets. His was a war waged against a twentieth-century enemy with the romantic ideals and courage of his Scots Highlander ancestry, reinforced with the Founding Fathers' revolutionary principles of political and economic liberty, and the religious principles of Calvin and Knox.

Congressman McDonald told a Marietta, Georgia, church audience shortly before his murder aboard KAL 007:

> We in this country have a rather unique heritage written into the basic documents of our government. We as Americans hold that we receive our fundamental rights, our basic rights, from a divine source, from God, not government. And it is the purpose of government to protect God-given rights; that statement

> is listed in the Mayflower Compact, in the agreement that drew up that little community in Jamestown, and was restated in the Declaration of Independence and in the Constitution of the United States.
>
> The Founding Fathers knew that any time a central government presents itself as being the provider of goods and services, two things inevitably take place. You study history, and one nation after another has fallen into this trap. First, you will produce a conflict society. Second, as sure as night follows day, the demands will always exceed the ability of the treasury to provide.[17]

At the conclusion of his lengthy talk, Congressman McDonald urged his audience not to endorse political parties and politicians who would not work to build a renewed appreciation for and put into practice the values of a limited government, free enterprise, and a living biblical morality. He asked his audience to see themselves as latter-day Signers of the Declaration of Independence and the Constitution who had risked their lives, fortunes, and sacred honor in defense of those ideas that had once made America great.

He concluded by quoting from the Book of Ezekiel in the Old Testament of the Bible; its forty-eight chapters warn against the threats to ancient Judah and Jerusalem. Ezekiel attempted to sustain his fellow exiles by seeking to keep alive their waning religious faith, fostering unity among them, and stressing the primary importance of taking responsibility for the promised restoration of Israel.

Congressman McDonald said:

> You may remember that in Ezekiel the Lord is making reference to the Watchman on the tower; if evil, villany, or war comes, and if the Watchman does not sound the trumpet, then the blood is on his hands. Any of the blood that is shed is on the hands of the Watchman who goes to sleep. But if the Watchman sounds the trumpet, and if the people are too lethargic, too busy enjoying the good life to care, then the blood and responsibility is on the hands of the people.[18]

The Rev. Joseph Morecraft III, a close personal friend of the Congressman, maintains that Dr. McDonald's long-range goal was to lead a Christian-based conservative reformation movement in America and the West. A movement based not on the submissive Christian doctrine dominant today that counsels surrender, appeasement, and even resignation to death, but one with a robust and affirmative philosophy of life in the tradition of Calvin, Knox, and the Founders of America, following the doctrine that it is the duty of every Christian to fight evil and tyranny.

Reverend Morecraft told this author, referring to the biblical significance of Dr. McDonald's death:

> One of the first things that God does when He begins to judge a nation is to remove effective leadership and replace it with leadership that has no commitment to the past and is only present-oriented and, therefore, reckless and irresponsible.[19]

When almost four thousand supporters of Congressman Lawrence Patton McDonald gathered at historic Constitution Hall on September 11, 1983, and shook its very rafters with righteous anger and indignation, it was the spirit of Knox and the Founding Fathers that moved through them and gave deeper meaning and expression to their loss. The sad mournful strains of the Donald Clan Scots bagpiper in the hall was a requiem for a warrior, poignantly reminding us of the long tradition of the Scots' refusal to submit to the sword of the tyrant without a fight.

The Congressman's brother, Dr. Harold McDonald, Jr., observed:

> The Constitution Hall memorial service was one of the most stirring ceremonies I have ever seen. I thought there would be two services, a more religious one at home and a political one in Washington. But the Washington one had as much a religious feeling as the religious one in Georgia. I had the feeling that, rather than being sad for Larry's loss, [those gathered] had a feeling that a soldier had died defending [them]. . . . You didn't have time to feel sorry for yourself when you were trying to stand up and be proud that you were the soldier's family.[20]

Dr. McDonald died as he had lived, a warrior secure in himself that what mattered most in this mortal existence was that we not remain passive spectators in the never-ending struggle to preserve and protect those values that make us more than a collection of cells and automatic reflexes.

In the final analysis McDonald's religious convictions gave him the incredible strength of character to endure in public life a remarkable amount of abuse, scorn, ridicule, and outright hatred. His refusal to reply in kind says as much about his own character as it does about those who preferred to believe the worst of someone who loved his country and his God as much as he loved life. What made Dr. McDonald different from other elected officials in the United States was that he had developed the capacity to allow the legacy of the living past to flow through him like a river, enriching the present and planting the potential for future greatness, while realizing that our singular existence in time and place is fundamentally spiritual.

JEFFREY ST. JOHN
January 1984

Appendix I

The following questions were submitted in writing to U.S. Secretary of State George P. Shultz on October 3, 1983. His written replies were made during the first week of November 1983. His responses to the author's questions appear unedited in the following question-and-answer format.

Q. When you first heard of the shoot-down of 007 what was your instant reaction?

A. Shock and revulsion.

Q. Was there concern in your mind, or that of your associates, that the 007 incident was a premeditated Soviet ploy to precipitate a challenge to the United States and bring on a super-power confrontation? In other words, was this ever viewed as a Soviet test of U.S. and Western allies' will?

A. No.

Q. Some Senators whom I have interviewed maintain that had the Japanese foreign minister not been so outspoken and vocal, the United States would have been less forceful in its initial response. How would you characterize U.S. intentions from the first hours?

A. After the Korean government had informed us the airliner was missing, our first concern was to locate it. Early in the morning of September 1 we confirmed that KAL 007 had in fact been shot down. Under Secretary of State Lawrence S.

Eagleburger called in the Soviet chargé at 9:30 A.M. to protest the Soviet attack on the plane and to demand a complete and rapid response from the Soviet Union to our inquiries in the matter. This was followed at 10:45 by my initial press conference in which I expressed our revulsion at this action and demanded a full Soviet explanation. I made the decision to have this press conference and to release the information we had, not to inflame opinion, but because I believed the American people and the world at large needed to know the truth about what the Soviets had done.

Q. The Japanese were reported to be reluctant to release the radio intercepts for fear of compromising intelligence sources. Was that a worry, and what went into the decision to release the transcripts? What would have happened if they had not been made public?

A. The United States and Japan released transcripts of the Soviet pilots' communications, as well as other pertinent information, to show clearly to the world that the Soviet Union had indeed shot down KAL 007. It should be remembered that for six days after the attack, the Soviets refused to admit they had shot down the airliner. They said it had left their airspace. They only admitted they had shot it down because we and the Japanese had presented the world irrefutable evidence of the Soviet misdeed. The transcript of the radio transmissions between the Korean airliner and the air traffic controllers at Tokyo's Narita airport provided by the Japanese government clearly indicated that KAL 007's pilot had no indication that he was flying off course or that he was being pursued by Soviet interceptors.

Q. Would you comment on reports in the media that you privately voiced the view initially that the downing of 007 was not worth going to war. Was there ever the fear that the incident might lead to war?

A. At no time was there concern that the incident would lead to war.

Q. Why was the Washington-Moscow hotline not used to talk directly to the Soviets? Or was it used? What direct contacts were made to the Soviets in those first few hours or the first day? There has been a suggestion by some that the United States was slow and hesitant in its reaction.

A. When Japanese search efforts failed to locate the downed airliner, and as the possibility that the aircraft had deviated into Soviet airspace became more apparent, the Korean government, which does not have diplomatic relations with the USSR, requested that the United States urgently pass a message to the Soviets asking if they had any information regarding the fate of the aircraft. The message was relayed by Assistant Secretary Burt on our behalf and on behalf of the Korean government shortly before midnight August 31 to Soviet Chargé d'Affaires Oleg Sokolov. At approximately the same time in Moscow, U.S. Embassy Chargé d'Affaires Warren Zimmerman asked the same questions of the Director of the USA Department at the Soviet Foreign Ministry, Aleksandr Bessmertnykh. Frequent diplomatic contacts over the KAL issue continued over the next several weeks in both Moscow and Washington. Why the Soviets didn't contact us urgently—over the hotline or by any other means—during the two and a half hours they held KAL 007 under observation and claim they thought it was a U.S. "spy plane" is a good question, and one you might put to the Soviets.

Q. At what point did you or your advisors conclude that the Soviets' shoot-down of 007 was deliberate and no accident?

A. While there is no conclusive evidence whether or not the Soviet pilot knew he was shooting down a passenger aircraft, there is no question that the Soviets deliberately shot down a foreign aircraft without making an adequate effort to warn it or identify it. Underscoring the deliberate nature of the Soviet action was Gromyko's subsequent warning that a future border violater would suffer the same fate. Soviet indifference as to whether an intruder is civilian or military was shown in 1978 when a KAL airliner was fired upon even after the Soviet interceptor pilot identified it as a passenger aircraft.

Q. Have you or your associates at State come to any firm conclusions as to what motivated the Soviets to shoot down 007 after tracking it for two and a half hours?

A. A reading of the pilots' conversations strongly suggests that the Soviets were bent on shooting down the aircraft before it escaped. Marshal Ogarkov in his September 9 press conference stated that the order to fire was given as the aircraft was over the hamlet Pravda, which is located on the edge of Sakhalin's western coast. The answer to the broader question of motivation seems to lie in the character of the Soviet Union. There is a massive concern for security, a massive paranoia, and I think this act was an expression of that excessive concern over security.

Q. What did the State Department do in conferring with its allies about the 007 incident?

A. From the very beginning of the incident, the United States and Japan worked in full cooperation. The United States, the Republic of Korea, and Japan immediately joined in calling for a meeting of the U.N. Security Council to discuss the Soviet attack. On September 7 and 8, I met with other NATO foreign ministers in Madrid to coordinate our response to the Soviet attack. On September 9, I met with foreign ministers of European neutral and nonaligned states attending the Madrid CSCE meeting. I reviewed the need for international action and requested their governments' support. At the same time in Washington, the State Department was consulting with a still broader set of countries, including those of Africa, Asia, the Middle East and Latin America. We cooperated with concerned states in presenting the Soviet Union with claims for reparations, as well as in taking action at the U.N. Security Council and at the International Civil Aviation Organization's (ICAO) extraordinary meeting September 15 in Montreal.

Q. When you met with Soviet Foreign Minister Gromyko, what conclusions did you draw from his attitude?

A. Gromyko's performance in Madrid—both before the CSCE participants and in his private meeting with me—made clear that the Soviet Union was determined to stonewall on this issue, and was not interested in finding a way to limit the damage this tragedy would cause to East-West relations.

Q. What do you make of the Soviet political leadership, for the first time in its history, using the Soviet military to deal with a firestorm of worldwide rage by holding that Moscow press conference?

A. It has not been unusual in recent years for the Soviet military to participate in press conferences with their civilian colleagues on matters with military relevance, such as arms control. The deputy foreign minister and the chief Central Committee propagandist were also present at the press conference and answered questions. This particular press conference attracted more than the usual attention due both to the presence of Marshal Ogarkov and the subject matter.

Q. The U.S. strategy was to make the case that 007 was the world against the Soviets, not the United States against the USSR. In view of the U.N. vote, did not that strategy of the U.S. fail?

A. No, on the contrary, I think it succeeded. The Soviet action was deplored by a firm majority of votes in the U.N. Security Council. Only the Soviets and Poland voted against the Dutch resolution deploring the destruction of the airliner. The ICAO Council and Assembly, by sweeping majorities which the Soviets could not veto, also deplored the Soviet action and called for an ICAO investigation which is now underway. The two-week boycott of air service between the Soviet Union and most NATO countries, as well as Switzerland and Japan, was an impressive and unprecedented demonstration of international unity which delivered a strong political message to Moscow.

Q. The U.S. strategy was keyed to galvanizing worldwide opinion against the Soviets. But what lasting effect can world opinion have on a regime that has in the past ignored and been contemptuous of such opinion?

A. The Soviets very much want to present a good image to the world. They were certainly stung by the public outcry against their action and will not want to repeat the experience again soon. What that may mean in terms of concrete action on their part is difficult to say. As the President said in his September 5 speech, "we know it will be hard to make a nation that rules its own people through force . . . cease using force against the rest of the world. But we must try."

Q. What led to the decision not to postpone arms reduction talks in view of the 007 incident?

A. In deciding that the arms control negotiations should proceed, President Reagan again reaffirmed the importance he attaches to the pursuit of effective arms control as an essential complement to our efforts to strengthen defense and deterrence.

Q. Why was it an inappropriate response to recall the U.S. Ambassador to Moscow and suspend various ongoing relations with the Soviets?

A. In formulating our response to this outrage, we considered a broad range of options. We believed that it was important that our reaction focus world attention, not on U.S.

retaliation, but on the Soviet Union's unacceptable and brutal act. As the President said on September 5, the Soviet action was a crime against the international community and called for an international response. The issue is the safety of international civil aviation. Therefore we decided to work with other concerned nations to coordinate a collective response in that sector, rather than concentrating on a series of unilateral steps.

Q. Can you characterize what real harm has been done to the Soviets, given the U.S. and allied response to the 007 incident?

A. The image the Soviets would like to project of themselves as a peace-loving country has certainly been dealt a severe blow.

Q. Whatever gain the United States and its allies made in galvanizing world public opinion against the Soviets, has not the 007 incident left the impression that the Soviets can act with impunity whenever they choose?

A. No, I don't think so.

Q. What did the 007 incident tell you about the Soviets?

A. Neither the action itself nor the Soviets' totally unacceptable response to the world community has surprised us. We have no illusions about Soviet international behavior. But for that very reason, our established policy towards the USSR, based on American strength, realism about Soviet aims and motives, and a willingness to talk about matters of mutual concern, has provided the appropriate framework for dealing with this crisis.

Q. The 007 incident has within a month disappeared from public discussion. What long-range consequences, if any, do you foresee its having?

A. I believe our firm and measured response to this abhorrent Soviet action, in conjunction with the international community, has demonstrated indeed that it is the USSR versus the world. At the same time, we intend to keep talking to the Soviets, to leave them in no doubt about our positions concerning not just the airliner but also other pressing international issues, and to encourage them to meet our urgent concerns as a responsible member of the international community. We believe the International Civil Aviation Organization (ICAO) investigation of this matter to be most important in ensuring such a tragedy does not recur, and we are encouraging other nations to support the investigation fully. We are also looking at measures within ICAO to strengthen civil aviation safety. One idea currently being considered is a French proposal to amend the Chicago Convention to ban the use of force against civil aircraft, subject to the provisions of the United Nations Charter. We are supporting the amendment, which will be considered by a special ICAO Assembly planned for early 1984. While this incident may no longer be on the front pages of the newspapers, it has sunk deeply into the consciousness of the world. It will not soon be forgotten.

Q. Finally, with the benefit of hindsight, what would you have done differently in the first few days of the crisis?

A. I believe our response was both appropriate and effective given the circumstances.

Appendix II

Proposed sanctions against the Soviets offered by Sen. Jesse Helms (R–N.C.) and reasons for opposition by the Reagan administration and Sen. Charles Percy (R–Ill.)

The following is taken directly from the Congressional Record *of the U.S. Senate debate on September 15, 1983, over the Joint House-Senate resolution condemning the Soviets' destruction of KAL 007. The excerpt contains a memorandum from the Reagan administration to the chairman of the Senate Committee on Foreign Relations, Mr. Percy, detailing its reasons for opposing eight specific amendments to the Joint Congressional Resolution. Senator Percy's remarks, before and after the memorandum was submitted for the record, are also included as representative of the liberal Republican and Democratic view of the KAL 007 tragedy, as well as the reasons for opposing specific sanctions against the Soviet Union, which the Reagan White House and State Department also opposed when they were offered by Senator Helms.*

MR. PERCY. Mr. President, there are many Members, including myself, who would share the emotions motivating Senator Helms in introducing an amendment to the resolution on the Korean airliner tragedy.

Certainly, those of us who have known Senator Helms know that he is a very tough-minded man. He has a tough mind but he has a very soft heart. He is a gentleman through and through.

When he spoke this morning on his feeling for those little children when they got on the airplane, recognizing that those children are treasures who will never be present on this Earth again because of the dastardly action of the Soviets, no one standing near him could question how deeply he feels about this. This is an emotional experience for so many people. I know that from when I suffered tragedy. When I first became chairman of the Foreign Relations Committee, I headed a delegation to South Korea for the inauguration of their President. We flew exactly over that same route. It could have been all the Members of Congress in that delegation.

I have great empathy for the feelings of my distinguished colleague. As I said before, I did have a number of amendments that I felt could be constructive to the resolution.

But I have come to the conclusion, following the passage by a unanimous House vote of 416 to nothing, and a breakfast meeting with the minority leader, Congressman Bob Michel, that it would be best to have no amendments. The House of Representatives does represent a full spectrum of views, ranging in every aspect and in every degree, but the House felt as a body that it was important to stand together on this and immediately send this message loud and clear and stand with the President so that there is no confusion about foreign policy. There tends to be in our form of society a sense of confusion, occasionally. It is the very nature of our society, that we do not always have one voice on issues. But here the world seems to be speaking with one voice.

It is clearly now the strategy of the administration, because it is in accordance with the feelings of the people of the world, that this is not just a U.S. issue. It is not the United States versus the USSR. It is not an East-West confrontation. It is the world, the whole world, that feels revulsion about this action.

And I pay particular tribute to the members of the Airline Pilots Association representing pilots around the world. They have taken steps and will I think follow through to see that the sanctions imposed against the Soviet Union to ground the largest civil aviation fleet in the world, Aeroflot, will be implemented in country after country. Our friends in Canada acted immediately in that regard, thereby making far more difficult the transit of military equipment from Moscow to Cuba. That action alone deprived them of landing rights in Gander and that means that there is at least an inconvenience to the Soviet

Union. It will be a humiliation as the world gradually brings to a halt the activities of that airline.

That is one thing that is already being done and I have suggested to the President a number of other things. But I do feel at this time it is highly desirable that we work together and act together.

Would that we could somehow simply ignore the Soviets, cut off trade, cut off negotiations, simply refuse to deal with them in any way, but that is not an option in our complex interrelated world.

The United States is a superpower. We cannot withdraw from our responsibilities to the Soviet Union to keep communications open at the highest level and to provide assurances to our friends who look to us to balance Soviet pressure and resist Soviet aims. If we attempt unilaterally to isolate the Soviet Union, we isolate ourselves and weaken our own ability to deal with our major adversary.

We must work with the international community. We must gain a consensus among our allies. We must not take measures which hurt ourselves more than the Soviet Union.

President Reagan's response to the Soviet destruction of Korean Air Lines Flight 007 has been appropriate and hard hitting. It has gained the acceptance of the international community. It has now gained the unanimous support of 416 House Members in the joint resolution passed yesterday. Let us not dilute the President's effectiveness. We want to support what the President is doing because he is doing it extraordinarily well.

I oppose amendments to the resolution because we have agreement on the leadership resolution. It was after all, sponsored by the majority and minority leaders; by the chairman of the Foreign Relations Committee and the ranking member of the Foreign Relations Committee; and by the chairman of the Armed Services Committee and the ranking minority member of the Armed Services Committee, Senators Baker, Byrd, Percy, Pell, Tower, and Nunn respectively.

I oppose amendments to the resolution because we do have extensive agreement. This resolution can be considered and pass the Senate today. It can go to the President today. And it will have his immediate signature.

Mr. President, the administration strongly opposes the Helms amendment. The Foreign Relations Committee has received specific comments from the administration this morning on the Helms amendment as it was introduced Tuesday. I would like to read those comments for the benefit of my colleagues because copies of these comments have not yet been received by every member of the Foreign Relations Committee.

Mr. President, I ask unanimous consent that just prior to the annunciation of the administration's provision on each of these several points of the Helms amendment that a copy of the original Helms amendment be incorporated which the administration is then responding to, taking into account that the Helms amendment has been modified, one item, at least, has been deleted from it. But these are the only comments I have as they relate to the original amendment.

THE PRESIDING OFFICER (Mrs. Hawkins). Without objection, it is so ordered.

The material follows:

HELMS: (1) recall the United States Ambassador to the Soviet Union for urgent consultations and reduce the number of Soviet diplomats accredited to the United States to the number of United States diplomats accredited to the Union of Soviet Socialist Republics;

ADMINISTRATION: 1. The United States Ambassador and United States diplomats accredited to the U.S.S.R. are in place vigorously to represent and to promote the interests of the United States. To reduce their number—or to invite such a reduction by reducing the number of Soviet diplomats in the United States—will diminish the capability of the United States to protect its citizens and promote its interests in the U.S.S.R.

HELMS: (2) conduct a comprehensive reappraisal of the complete spectrum of United States-Soviet relations, including arms control, human rights, East-West trade, and regional issues;

ADMINISTRATION: 2. As the President stated in his speech of September 5, the Administration has conducted long meetings, including with the Congressional leadership. The President stated that we cannot "give up our effort to reduce the arsenals of destructive weapons threatening the world." Thus he has spoken to arms control. We continue to promote our human rights concerns, East-West trade and regional issues.

HELMS: (3) report to the Congress on the record of Soviet compliance or noncompliance with the letter and spirit of all existing strategic arms limitation talks (SALT) agreements and other arms control agreements to which the Soviet Union is a party;

ADMINISTRATION: 3. Procedures exist for dealing with the Intelligence Committees of both Houses on the questions of compliance with arms control agreements.

HELMS: (4) direct the United States negotiators at the strategic arms reduction talks at Geneva to link the possible success of such talks with the willingness of the Soviet Union to abide by international law as a responsible member of the community of nations, paying specific attention to the KAL 7 massacre, Soviet violations of the Helsinki accords, the Soviet invasion and subjugation of Afghanistan, the repression of Poland and its free labor movement, and the use of chemical and biological weapons in contravention of existing treaties;

ADMINISTRATION: 4. With regard to linkage of the START talks with Soviet actions, including the KAL massacre, as noted above the President stated that we must not give up efforts to reduce the arsenals of destructive weapons.

HELMS: (5) reemphasize the inconsistency of the Soviet military presence in the Western Hemisphere with the Monroe Doctrine;

ADMINISTRATION: 5. The President and other Administration officials have stated on many occasions and we have communicated to the Soviets our dissatisfaction with their activities in the Western Hemisphere.

HELMS: (6) declare Poland in default on all or a part of the debt owed to the Commodity Credit Corporation, recognizing that Poland is an integral part of the Soviet economic empire and that financial credit is an element of national strategy;

ADMINISTRATION: (6) Default would take the heat off Poland to continue to pay at least some portion of what it owes.

Default would not stop credits from going to Poland as some allege; no one is lending to Poland now anyway.

Poland has few assets in the West which could be attached in

event of default. Thus, default is one sure way not to get paid.
Poland would have more cash if default were declared and would certainly need less short-term financial help from USSR.

———

HELMS: (7) tighten substantially the foreign policy and military controls over the export of machine tools, high technology products, and equipment for the development of Soviet oil and gas resources; and

ADMINISTRATION: (7) The President stated in his September 5 speech that we are redoubling our efforts with our allies to end the flow of military and strategic items to the Soviet Union.

———

HELMS: (8) direct the Secretary of the Treasury to use existing statutory authority to prevent the import of any product or material produced in the Soviet Union unless the President certifies that it was produced without the use of forced labor.

ADMINISTRATION: (8) The President and all Americans abhor Soviet use of forced labor. Should it be demonstrated that products or materials produced by forced labor are being imported into the United States the Administration will take appropriate steps.

MR. PERCY. Mr. President, I understand the desire of the senior Senator from North Carolina to include in the resolution specific recommendations to the President. I would like to take just one point that is made in the revised Helms resolution, which I read as follows:

> It is the sense of the Congress that the President should direct the U.S. negotiators at the strategic arms reduction talks at Geneva to link the possible success of such talks with the willingness of the Soviet Union to abide by international law as a responsible member of the community of nations, paying specific attention to the KAL 007 massacre, Soviet violations of the Helsinki accords, the Soviet invasion and subjugation of Afghanistan, the repression of Poland and its free labor movement, and the use of chemical and biological weapons in contravention of existing treaties.

A strict interpretation of that, for instance, reestablishment, let us say, of the free labor movement and labor unions in Poland, just to take one example, literally means that in the course of the Reagan administration's first term, and its expected second term in office, there would simply be no arms control agreement of any kind with the Soviet Union.

If it is a condition precedent to our willingness to enter into such agreement that there is definite linkage between any of these activities, abhorrent as they are, then I think we are simply saying we are going to shelve arms control. The President has already made his decision to do just the contrary. We would be in a position where we are tricking the President not to do something that he has already declared he will do to the world and to our citizenry, that we will carry on arms control negotiations despite our abhorrence of this tragic event.

I am not now speaking on behalf of the administration. I am describing my own strong feelings about this matter. Ambassadors Rowny and Nitze strongly support the President's viewpoint that we should move forward with these talks. Both men are admired tremendously by every single Member who is sponsoring this resolution, I believe. Arms control is not done as a favor to the Soviet Union. It is in our national interest to move the superpowers back from the brink of a nuclear holocaust.

The President has emphasized that the shooting down of KAL flight 007 is an issue of the Soviet Union versus the world. The world has vital interests in assuring that civil air travel is made safe. But there are 4.7 billion human beings living on this planet. They have even a greater stake in making sure that life on Earth is not ended by a nuclear war.

I go back now 23 years to the inaugural address of John F. Kennedy, who noted in that address that we have the power to eradicate all hunger and poverty throughout the world. Then he added that we have the power also to eliminate all human life.

Now, some 50,000 nuclear weapons later, he could have amended that and said, if he were living today, we have developed the capacity to eradicate every living thing on this planet, with just the exchange of the existing weapons, weapons that exist today on the part of the Soviet Union, the United States and our allies, as well as those that are possessed by the People's Republic of China.

There is no question that life would be extinguished, perhaps for centuries to come.

The Korean Air Lines disaster shows us that the Soviet military is prone to shoot first and ask questions later when they see targets come toward them on radar. It is debatable whether they ever really tried to confirm what they were shooting at before they fired their missiles at the airliner.

We can only imagine and we now have a basis for fearing—the whole world has—how the Soviets might react if they thought their radar showed they were under a nuclear missile attack. It is ironic that we are, right now, trying to engage the Soviets in START in a discussion of various confidence-building measures, one of which would allow each country's military command post to communicate directly if they thought an attack was occurring.

I have suggested earlier today, and I shall expand at some point on the floor of the Senate later the details as to how this can be accomplished, that we might take a global satellite system, which will be in place by 1988, and expand it to civil aviation.

What would happen is that those START talks would be suspended. Those talks are being carried on now and arrangements are being made so that we can reduce the possibility that those signals will come over the black box carried by the President of the United States or his aide every place he goes. The gravity of the problem struck me anew earlier this year. In an elevator in the Conrad Hilton hotel on January 19, there were the four of us, the military aide, the President, my wife, and myself. If that signal had come in, he would need every sort of communication capability possible.

What we are trying to do is prevent destruction of all humankind, all life on Earth.

As I say, the Soviets have shown a habit and have now demonstrated again that they shoot first and ask questions later, when they see targets coming toward them on radar. If this amendment were to carry, we would be directing the President to suspend START talks and thereby lose the opportunity to continue negotiations on this much-needed confidence-building measure. And this is only one example.

As I said before, I do understand the desire of the senior

Senator from North Carolina and his cosponsors to include in the resolution specific additional recommendations of the President.

(Mrs. Hawkins assumed the chair.)

MR. SYMMS. Madam President, will the Senator yield on that last point?

MR. PERCY. I should have to yield on the Senator's time, I am afraid. I may run out of time.

How much time do I have?

THE PRESIDING OFFICER. The Senator has 2½ minutes.

MR. PERCY. I shall probably need that amount of time.

Each of us has his or her own idea of ways to emphasize our Nation's condemnation of the Soviet attack on the unarmed civilian airliner. We do have this House-passed resolution before us which expresses our outrage very clearly and which carried in the House by the impressive, overwhelming margin of 416 to 0.

At this time, I submit that it is highly desirable to put our individual ideas aside and confirm the House language, which is almost identical to the Senate leadership resolution, so that the resolution can go to the White House for the President's signature today. If we accept a single amendment, we shall have to go to conference with the House and precious time will be lost. We want to confirm the House action today, which was to show the world that the President, the House of Representatives, and the Senate stand united with the same words with one voice in condemning the Soviet Union for its wanton act against international civil aviation and against humanity itself.

For these reasons, Madam President, I urge my colleagues to vote against the amendments before us. Let us support the President in this crisis. Let us adopt the House-passed language so that the resolution can go to the President today.

MR. RANDOLPH addressed the Chair.

MR. PERCY. Madam President, I believe the Senator from West Virginia wishes to speak against the pending amendment: is that correct?

MR. RANDOLPH. I wish to make only approximately 50 to 60

seconds of comment. I would rather not make it off the bill at this time. I want to be cooperative.

MR. PERCY. I yield whatever time I have remaining in that case.

Appendix III

The following is a complete list of the passengers and crew who perished aboard Korean Airlines flight 007 on September 1, 1983, over Soviet-occupied Sakhalin Island. It is taken from Massacre in the Sky—The Soviet Downing of a KAL Passenger Plane, *published in September 1983 by the Korean Overseas Information Service, Seoul, Korea.*

Passengers

Korea (75 persons). Kwon Yon-kum (f), 72; Yu Kyong-gun, 27; Wi Kang-il, 40; Pak Heung-sol, 64; Lee Chae-il (f), 52, Pak's wife; Pak Hong-sun (f), 45; Chong Hwa-sun (f), 30; Lee Chun-hyok, 6, Mrs. Chong's first son; Lee Chun-won, 4, Mrs. Chong's second son; Chang Yong-tae, 44; Kim Kum-sun (f), 45, Chang's wife; Chang Sung-nyong, 16, Chang's first son; Chang Sung-jip, 12, Chang's second son; Yu Kap-il, 43; Ko Yong-ho, 48; Pak Han-tae, 42; Cho Chae-muk, 31; Lee Chol-gyu, 30; Lim Won-bok (f), 60; Min Kyong-hun, 26; Sohn So-ja (f), 42; Sohn Yong-ja (f), 40; Un Myo-sun (f), 65; Pak Song-ha, 31; Chung Yom-sun (f),49; Pak Min-sik, 35; Kim Ae-gyong (f), 33, Pak's wife; Yu Tong-yol, 53; Pak Chang-on, 41; Lee In-ho, 24; Lee Yon-pyo, 44; Lee Myong-jae, 30; Kim Sun-taek, 40; Kim Rae-su, 49; Paek Yun-jong, 36; Kim I-gyu, 45; Pae Pun-sun (f), 68; Yun Song-bu, 30; Kim I-sik, 26; Kim Chin-hong, 43; Han Ung-jon, 37; Yu Ok-myong, 33, Han's wife; Han Man-chol, 7, Han's son; Han Chong-min, 4, Han's daughter; Kim U-sik, 66; Yu Pyong-suk, 66, Kim's wife; Kim Yong-sik, 36; Lee Un-hyong, 37; Chong Mu-ho, 34; Kim Pom-chon, 29; Yu Chun-taek, 39; Kim Kan-nan (f), 72; Yun Ae-sik, 22; Lee Hi-yong, 38; Choe Kyong-ae (f), 38, Lee's wife; Lee Song-jun, 13, Lee's son; Lee Kw-hyon (f), 10, Lee's daughter; Han Sonsok, 48; Chang Sang-jun, 71; Hwang Pyong-suk (f), 63, Chang's wife; Chae Su-myong, 12; Kim Chong-yung, 43; Oh Chong-ju (f), 52; Lee Chong-bong, 37; Lee Sang-gyun, 27; Kang Yong-chae, 9; Lee Sun-sik, 37; Pak Il-chong, 40; So Chu-ok (f), 29; Lee Chung-gun, 4, So's son; Lee Myong-hwan, 31; Lim Chong-chol, 50; Kwon Song-hi, 19; Kim Hyon-bok, 37; Chong Ok-sun (f), 45.

Japan (28 persons). Ishihara Masuyo (f), 63; Haba Hiroki, 18; Kobayashi Ikuko (f), 57; Kobayashi Shoichi, 60; Kono Tomiko (f), 31; Kitao Hitomi (f), 29; Kawana Hiroaki, 20; Inoue, K (f), 39; Inoue, M (f), 13; Inoue, A, 3; Mano Sayori (f), 22; Osedo Midori (f), 63; Shiiki Shizue (f), 43; Shiiki Lune (f), 12; Nakazawa Takeshi, 25; Tomaguchi Mikio, 34; Tomitaka Yaeko (f), 55; Tanaka Keiko (f), 35; Takemoto Kiyonori, 21; Takemoto Tomiko (f), 47; Osaka Noriyuki, 39; Shimizu Miyako (f), 52; Yoden Kazuko (f), 28; Yamaguchi Masakazu, 32; Okai Makoto, 22; Okai Yoko (f), 24; Nakao, Osami, 24; Stevens Hiroko (f), 26.

USA (61 persons). Carrasco, C., 14; Carrasco, M.E. (f), 18; Guevara, T., 1 month; Chouapoco, M. (f), 34; Chouapoco, C. (f), 4; Forman, E. (f); Ocampo, M. (f); Ocampo, C. (f), 3; McDonald, Larry, 48; Orwen, William, 32; Benis, Richard, 32; Campbell, S. (f), 28; McNiff, Kevin, in his twenties; Ephraimsonabt, A. (f), 23; Lombart, Don; Lombart, Aiden; Chambers, Joyce (f), 34; Dawson, Lucille (f), 55; Fitzpatrick, Lillian (f), 60; Slaton, Jessie (f); Zarif, Margaret (f); Steckler, S., 32; Steckler, I. (f); Chouapoco, C., 55; Kohn, Allan, 63; Kohn, Lilian (f), 54; Bessell, Eleanor (f); Culp, Marie (f); James, Hazel (f); Kole, Muriel (f); McGetrick, Mark, 29; Miller, Edna (f); Swift, Frances (f), Ariyadez, Diane (f), 28; Ariyadez, S. (inf.), 8 months; Dorman, S., 55; Beirn, J., 40; Petroski, R., 38; Scruton, R. (f); Metcalf, Chong (f), 30; Metcalf, Rita (f), 7; Metcalf, Christa (f), 3; Truppin, Michael; Zareh, D., 45; Brownspier, K. (f), 35; Weng, M.T., 42; Oldham, John, 30; Bayona, Lilia (f), 35; Bayona, Anita (f), 29; Burgess, James; Katz, Jack, 62; Draughin, S. (f); Ellgen, R.; Wuduun, S. (f), 21; Chan, Joseph, 28; Song Ahn-na (f), 62; Hong Hyung-woong, 39; Pak Sa-ra (f) 5; Pak Gra-han, 3; Yoo Jung-soo, 47; Lim Jong-jin, 63.

Philippines (15 persons). Bolante, E., 60; Guevara, A. (f),

30; Cruz, E., 51; Cruz, F. (f), 51; Caser, C., 52; Lantin, Raymundo, 36; Ocampo, S.E. (f), 40; Putong, J. (f), 69; Omblero, A., 25; Galang, B., 30; Avecilla, A. (f), 26; Cruz E. (f), 23; Cruz, A., 50; Bolante N. (f), 23; Chan Amado, 30.

China(R) (23 persons). Cheng Chin Chong, 29; Chan Su Jen, 63; Chanlin Yee Shine (f), 57; Chen Ju Yen (f), 29; Tien Chi (f), 62; Pan Limei (f), 24; Liu Yin Chien, 32; Kong King, 52; Chang Mason, 25; Tsao Yuen Che, 38; Chen Fulong, 36; Kung Ching Fen, 60; Lin San Mei, 36; Chang Tsai Chen, 63; Yeh Ching Liu, 38; Wang Yun Sheng (f), 54; Ma She Jen, 50; Chen Lee Jenrong (f), 32; Chen Shiaofen (f), 7; Liu Chong Yun Shin (f), 33; Liu Po Shen (f), 10; Liu Chao Fu, 9; Lee Chi Cheng, 57.

Hong Kong (12 persons). Ho Yuk Yee (f), 28; Leung Ko, 56; Lai Yung (f), 49; Ho Ming Tai, 27; Lim, S., 38; Lui, John, 21; Yuen Chi Bong, 45; Yuen Wai Sum (f), 9; Yeung Oi King (f), 40; Siu, Robin, 35; Wong, Michael, 28; Iu Waikong, 32.

Canada (9 persons). Yeh, C.L. (f), 80; Covey, Merryloum (f), 34; Gregoire, John Paul, 65; Hendrie, M. (f), 25; Demassy, F., 26; Robert, F., 26; Panagopoulos, G., 32; Sayers, L., 24; Leung Chi Man (f), 47.

Thailand (6 persons). Sripoon, J. (f), 28; Hansuwanpisit, A. (f), 35; Homlaor, T., 8; Pakaranodom, S., 38; Pakaranodom, W. (f), 37; Pakaranodom, S. (f), 8.

Australia (5 persons). Grenfell, Neil, 37; Grenfell, Carol Ann (f), 34; Grenfell, N. (f), 5; Grenfell, S. (f), 3; Moline, Jan (f), 45.

Sweden. Hjalmarsson, 38.

Malaysia. Siow Woonkwang, 23.

India. Patel, K., 30.

Dominica. Nassief Authony, 40.

Britain. Powrie, Ian Dun, 25.

Vietnam. Dong Lochun, 20.

Crew

Chun Byung-in, 45, captain; Son Dong-hui, 47, copilot; Kim Ui-dong, 32, flight engineer; Kim Suh-il, 52, captain on flight to USA; Byun Hyun-mok, 52, copilot on flight to USA;

Lee Yoon-jae, 37, flight engineer on flight to USA; Ahn Insoo, 49, captain on flight to USA; Kim Hee-chul, 44, copilot on flight to USA; Kang Jin-hae, 44, flight engineer on flight to USA; Yoon Yang-no, 34, purser; Kim Hak-yoon, 30, assistant purser; Lim Sang-ki, 27, steward; Roh Jun-sik, 33, steward; Roh Kwang-woo, 30, steward; Kim Choong-yul, 35, steward; Sin Jung-moo, 42, steward; Chung Kum-joo (f), 27, stewardess; Park Yoon-sung (f), 25, stewardess; Kim Kyung-seung (f), 24, stewardess; Lee Un-hui (f), 24, stewardess; Kim Youngnan (f), 22, stewardess; Kang Boo-ye (f), 27, stewardess; Lee Un-mi (f), 27, stewardess; Suh Jung-suk (f), 24, stewardess; Bae Young-rang (f), 25, stewardess; Cho Kyung-ja (f), 22, stewardess; Kim Mi-hyang (f), 20, stewardess; Cho Hyangsim (f), 23, stewardess; Chung Ran (f), 22, stewardess.

Appendix IV

The Constitution and the Congress:
The dissenting views of a House Democrat

The following interview with Congressman Lawrence P. McDonald was conducted by the author on June 2, 1982, at the Cannon House Office Building, U.S. Capitol, Washington, D.C. It was prepared and edited originally for publication in the question-and-answer section of the Washington Times, *for which the author was a political correspondent. It was, however, subsequently rejected as unsuitable for publication since its subject was considered not in the mainstream of congressional thought.*

It is published here for the first time, revealing the untold, behind-the-scenes story of the Democratic House leadership's secret effort to overturn the results of the 1980 election.

In June of 1982 the U.S. Congress was deadlocked over budget and tax policy. The blame for this congressional anarchy was laid at the door of the Reagan White House by the House Democratic leadership. However, Rep. Larry McDonald (D–Ga.) charges in the following interview with *Washington Times* correspondent Jeffrey St. John that the responsibility for the continuing congressional deadlock really rests with the Democratic House leadership. McDonald maintains the months of congressional chaos and confusion are the consequences of a New Deal-liberal Democratic leadership minority's dictating to the moderate conservative House majority. He further maintains that such a liberal minority has created and benefited from what he terms "our conflict-ridden society."

Q. Why has the Democratic House leadership, supposedly in the majority, been unable to agree on a budget or a tax policy?

A. First, the House Democrats are not monolithic in their viewpoint. But you are beginning to have a resurgence of traditional Democrats, more constitutional, more Jeffersonian than New Deal Democrats. In short, a realization is growing in the country that Republicans are not just the answer, and you are finding opportunities for Conservative victories in the Democratic party. After the 1980 elections it was clear what the country wanted. What they didn't want were big spenders, big government advocates. Now fearing this change in the mood of the country and fearing this transition, Democratic House leaders, composed of liberal New Deal types, dishonestly tried to further their power so they could continue business as usual in defiance of the country and the new crop of congressmen sent to the House.

Q. You used the term "dishonestly tried to further their power." How did the Democratic House leadership further its power and to whom are you referring?

A. We are talking about House Speaker Thomas "Tip" O'Neill; we are talking about House Majority Leader Jim Wright. They were the principal architects of a plan for maintaining their power and control over critical House committees to prevent a historic transition if not an end to the New Deal, Fair Deal, Great Society idea that the principal function of Congress and the federal government was to redistribute income.

The principal plan was to stack and load three critical committees: Appropriations, Rules, and Ways and Means. The plan was passed to give a dishonest number of seats on these three key committees to Democrats over Republicans. This was accomplished by and large with a partisan vote that took place last year [1981]. At the time threats were made by the House leadership that punishment would be forthcoming for those who didn't go along. Those committees were heavily stacked with Democrats who, by and large, believe we should continue to use government as an instrument for forced redistribution of income. Now these three stacked committees did not reflect numerically or philosophically the makeup of the majority of the House. Now, the more conservative Democrats said this was wrong, if not dishonest.

Q. Did this stacking of the three key committees in the House last year play any part or contribute to the confusion, chaos, and paralysis we have seen all this year?

A. I think so, absolutely. The result of stacking these three House committees has been bills that philosophically do not reflect the general makeup of the House, nor do they reflect the thinking of the American people. What appears to be a House that is confused and divided and blamed the Reagan administration and special interest groups is really the results of this stacking process by the House leadership. I had to laugh when Speaker O'Neill on the House floor earlier this year was pleading that we must not fall apart in party unity. Almost with tears in his eyes he made this plea for unity, when he was the architect of this chaos and confusion. He was the architect for presetting the dials of the machine in the House that guaranteed chaotic and confused results. Majority Leader Jim Wright got a great deal of mileage out of his statements that the chaos and confusion in the Congress this year indicate that America is becoming fragmented, Balkanized. He said this was a dangerous development. Well, he and Speaker O'Neill and other so-called Democratic leaders in the House are responsible for this dangerous development.

Q. It has always been a cardinal rule of American politics that you don't raise taxes in an election year. How is it that both Democrats and Republicans are proposing to break that cardinal rule in 1982?

A. Logically, you don't raise taxes in an election year. But I think the Democratic leadership decided very early to set out to discredit Reaganomics in both the broad and narrow sense, to discredit the program by advocating tax increases as a way to fight the recession blamed on Reagan and the Republicans. The excuse for tax increases was the high deficits that got a great many unthinking Republicans to go along, who couldn't see what the Democrats were actually doing in advocating a tax increase. It wasn't to reduce the deficits; it was to discredit Reaganomics.

Q. But you can't blame Mr. Reagan's difficulties solely on the political designs of his Democratic liberal opponents, can you?

A. No, not entirely. The fundamental problem with President Reagan is that he is not a hardball player when it comes to politics. He's very persuasive. But I have come to the conclusion he's the type of person who likes to sit around a table for a friendly chat, work out any disagreements and try and resolve them within an hour and a half, work out an agreement, reach a general compromise, everybody smiles, shakes hands, and then he says, "Let's go to dinner." This is a mistake because it's not hardball politics, which essentially identifies the source of the problem, who is the real enemy. In this case it's Congress. It has, in fact, been the problem for fifty years or longer. Sixty percent of the Federal budget is devoted to forced redistribution of income. Now the Founding Fathers and framers of the U.S. Constitution worked to prevent government from becoming what it is today. So in the most realistic sense the U.S. Constitution as far as Congress and the courts are concerned is nothing but a dead-letter document.

Q. But you regard yourself as a Constitutional Jeffersonian Democrat. Yet, what you are saying is that as we approach the two hundredth anniversary of its adoption in 1987 that it is no longer operative. Isn't that what you are saying?

A. Absolutely. This erosion of the authority of the U.S. Constitution to check the power of elected officials has been going on for most of this century, if not earlier. And as the Founders and Framers predicted, with this erosion of a check on the powers of politicians would come the society of conflict, not consent, not social, political, and economic concord.

Q. How does this historical erosion of constitutional checks on Congress and on the executive and judicial branches of the federal government relate to what we are now seeing in the chaos in Congress?

A. It's a manifestation of this lack of restraint by the Constitution on the Congress. When this has happened in the past conflict always follows as night follows day. As the limits upon government have broken down, as these limits have been destroyed, the federal government in essence may do anything that Congress says it thinks it ought to do. The federal government today, thanks to both the Congress and the White House for most of this century, is involved in almost every sphere of human endeavor. Not only in this country but around the world. As a result, conflict is the result of various people who want people's taxes to be used for this project or that project. In the final analysis, these competing groups, which politicians in Congress cultivate, must clash and create conflict.

Q. So it's not entirely Congress's fault that we have what you call a conflict-ridden society?

A. No, not entirely. The federal judiciary has broken down; it has also been an activist in the destruction of the Constitution. Congress is somewhat the creature of the people. In the final analysis, the problem is in the living rooms of America. We have lost an informed electorate. Which leads me to the most powerful branch of government. The national news media now has the power to drum up interest in any issue and present it as "news." They, along with Congress and the judiciary, have contributed to using government as an instrument for social engineering that has led to a conflict-ridden society as various groups compete for a piece of the federal pie. The news media play an important role in creating conflict and in destroying the Constitution by editorializing in the news and creating the appearance that mass support exists for various social programs.

Q. If we are, as you claim, more a society of conflicting groups than a society governed by a limiting document like the U.S. Constitution, what kind of society are we heading toward as we prepare to celebrate the bicentennial of the U.S. Constitution in 1986–87?

A. Well, we only have to consult the Founding Fathers and the Framers to answer that question. They were very worried about "faction" as a deadly element that destroyed Greece, Rome, and other ancient republics who wanted security more than they wanted liberty. So the very conflict created by the Democratic House leaders, which they regularly decry as dangerous, can in the final analysis only lead to a disorder and bankruptcy. The road to the American Revolution and the French Revolution was paved when power was abused and pulled to the center of both societies.

Certainly Rome offers the most famous and enduring warnings for our time as it did for the Founders and Framers. The fragmentation of Roman society was the result of playing off one group against another, resulting in conflict. Cicero probably represents one of the last voices of the Roman republic before the coming of the Caesars. And I am afraid we are moving in America toward the Age of the Caesars. Frankly, we've got just the remnants of our republic left. We may pledge allegiance to the flag and "the republic for which it stands," but I seriously doubt whether one person in a hundred today knows what a republic is. Or for that matter, that it is not a republic that we have today.

Q. The Tip O'Neills and the Jim Wrights, are they the New Caesars?

A. Only in part. The real coming of the Caesars was after the chaos. We are moving into this period of chaos and conflict.

Right now we are in the age of the demagogues who are leading us down the road to chaos and conflict. The Caesars appeared as saviors from the chaos. Certainly you can find more recent parallels in history than just Rome. The French Revolution ended with a savior and Caesar, Napoleon. In this century in Weimar, Germany, prior to the coming of a Caesar savior, was Hitler. The runaway inflation caused by government spending in Germany in the 1920s wiped out stability and made possible a Hitler.

We are producing a conflict-ridden society by the same process that destroyed past nations and civilizations. We have a fragmentation process going on because we have allowed the slow destruction of the U.S. Constitution. And the so-called House leadership of my party has been in the vanguard of the destruction of the U.S. Constitution as a limiting document. It's not a very nice or noble present for the American people for the two hundredth birthday of the Constitution in just four short years!

About the Author

Jeffrey St. John, with the publication of the *Day of the Cobra,* completes twenty-five years as a professional journalist, author, and broadcast news commentator and producer.

He is the author of the Time-Life book *Noble Metals,* published in the spring of 1984, and two other books. He is currently the editor-publisher of *The United States Times* and writes a syndicated column, "Headlines and History," relating significant events in American history to contemporary affairs. The feature was heard worldwide on the Voice of America and translated into twenty-six foreign languages during the years 1982–83. "Headlines and History" is currently heard nationwide as a syndicated radio feature. St. John is also a director of the U.S. Motion Picture Institute, Hollywood, California.

Mr. St. John has been a broadcast commentator for the CBS Radio-TV "Spectrum" series and for five years was a regular broadcast news commentator on the Mutual Radio Network where he was also the moderator for the weekly news-interview program, "Reporter's Roundup." He was for a year the NBC-TV "Today" Show business correspondent, and was an independent producer-moderator for New York TV stations; he conducted a regular broadcast talk show on WMCA, New York City, and later on WRC-Radio, Washington, D.C.

He is the winner of two Emmys from the National Academy of Television Arts & Sciences, one for a public affairs program on WTTG-TV, Washington, D.C., and one for a full-length Westinghouse TV documentary. He is also the recipient of the George Washington Medal of Freedom, Freedoms Foundation, Valley Forge, Pennsylvania.

For five years he wrote a syndicated column for the Copley News Service, and over the years his articles have appeared in the *New York Times, Washington Times, Philadelphia Inquirer, Chicago Tribune, Wall Street Journal,* and *Nation's Business.*

Born in Philadelphia, Pennsylvania, on July 14, 1930, he attended Louisiana State University before serving as a combat correspondent with the U.S. Marine Corps during the Korean War. Beginning during the Eisenhower years in the late 1950s, he has been a White House and Capitol Hill correspondent, and as a journalist, news commentator and columnist-author has covered every president since then. His assignments overseas have taken him to Europe, the Middle East, the Far East, and Latin America. Since 1977 he has made numerous personal fact-finding trips to Central America.

He is married to the former Kathryn Boggs, and they live in historic Southside Virginia. A dog lover, he raises Doberman Pinschers.

Notes

Introduction

1. A. L. Rowse, *Appeasement: A Study in Political Decline, 1933–39* (New York: W. W. Norton, 1961), p. 118.
2. Telford Taylor, *Munich: The Price of Peace* (New York: Random House, 1980), p. 1004.
3. Gen. Lewis W. Walt, USMC (Ret.) and Gen. George S. Patton, U.S. Army (Ret.), "The Swiss Report, a Special Study for Western Goals Foundation," Letter from the chairman, Lawrence P. McDonald, Western Goals, Alexandria, Va., Mar. 1983.
4. Lawrence Patton McDonald, conversation with author, London, Mar. 4, 1980.
5. McDonald, *We Hold These Truths* (Seal Beach, Calif.: '76 Press, 1976), p. 13.
6. McDonald, "The War Called Peace, The Soviet Peace Offensive," Letter from the chairman, Western Goals, Alexandria, Va., May 1982.
7. "No Place to Hide, the Strategy and Tactics of Terrorism," a transcript of a television documentary, interview by Congressman McDonald (D–Ga.), Western Goals, Alexandria, Va., 1982, pp. 13–14.

Chapter 1

1. Max Eastman, *Reflections on the Failure of Socialism* (New York: Devin-Adair, 1955), p. 57.
2. *Congressional Record,* House, Sept. 14, 1983, House Resolution 353, p. H6863.
3. *Congressional Record,* Senate, Sept. 15, 1983, Senate Joint Resolution 158, p. S12259.
4. Text, the White House, Office of the Press Secretary, for Immediate Release, Sept. 5, 1983, "Address of the President to the Nation," the Oval Office, 8:00 P.M. EDT, pp. 1, 3.
5. Text, the White House, Office of the Press Secretary, for Immediate Release, Sept. 17, 1983. Radio Address of the president to the nation, Camp David, 12:03 P.M. EDT.
6. Text, the White House, Office of the Press Secretary, for Immediate Release, Sept. 3, 1983, Photo Opportunity with the president and Ambassador Paul Nitze, the Rose Garden, 1:15 P.M. EDT, p. 1.
7. Ibid.
8. *Congressional Record,* Senate, Sept. 15, 1983, p. S12340.
9. Sen. Orrin Hatch (R–Utah), Text of speech, "What Should the Friends of South Korea Be Doing?" Skulla Hotel,Seoul, South Korea, Sept. 3, 1983, p. 1.
10. Ibid., p. 2.
11. Ibid.

12. A. L. Rowse, *Appeasement: A Study in Political Decline, 1933–39* (New York: W. W. Norton, 196l), p. 117.

13. *Congressional Record*, House, Sept. 14, 1983, p. H6884-85.

14. McDonald, "Technology Transfer to Soviets,"column for release, Aug. 23, 1983, Information-News Report.

15. ———, "Technology and Our Enemies,"column for release, June 14, 1983, Information-News Report.

16. *Congressional Record*, House, Sept. 14, 1983, p. H6867.

17. Richard E. Scott, "U.S. Businessmen Plan Moscow Fair Despite 007 Incident," *Washington Times*, Sept. 23, 1983, p. 3A.

18. McDonald, transcript "Reporter's Roundup" news interview, Mutual Broadcasting System, recorded July 21, 1977, House Radio-TV Gallery, U.S. Capitol, Washington, D.C. Broadcast over Mutual Radio Network, Aug. 7, 1977, p. 12.

Chapter 2

1. Seymour Weiss, "What If the Soviets Mean Every Deed of It?" *Wall Street Journal*, Sept. 23, 1983.

2. *U.S. News & World Report* Chronology, "Airborne Aggression: A Soviet Trademark," Sept. 12, 1983.

3. Library of Congress, "Soviet Downing of a South Korean Airliner," Analysts & Specialists, Foreign Affairs and National Defense Division.

4. Oleg Penkovskiy, *The Penkovskiy Papers* (New York: Avon Books, 1966), p. 355.

5. Frederick Barghoorn, *Soviet Foreign Propaganda* (Princeton, N. J.: Princeton University Press, 1964), p. 114.

6. "The Rules of the Game," *Time*, Sept. 12, 1983.

7. "Why the Russians Did It," *Newsweek*, Sept. 19, 1983.

8. "Rules of the Game," *Time*.

9. Charles M. Lichenstein, Deputy U. S. Ambassador to the U. N. Security Council, full text, U.S. Mission to the United Nations, statement in the Security Council, on the complaint of the United States and the Republic of Korea, Canada, and Japan, Sept. 2, 1983, p. 4.

10. Bruce Herbert, interview with author, Center for International Security, Washington, D.C., Sept. 23, 1983.

11. "Turning on the Heat," *Time*, Sept. 19, 1983.

12. *Congressional Record*, Senate, Sept. 14, 1983, p. S12317.

13. Ibid., p. 12316.

14. Ibid.

15. Larry Campbell, "Air Defense System Needs Beefing Up," *Anchorage Daily News*, Sept. 10, 1983.

16. Ibid.

17. Weiss, "What If the Soviets?"

18. *Congressional Record*, House, Sept. 12, 1983, p. H7106.

Chapter 3

1. *Congressional Record*, House, Sept. 14, 1983, p. H6876.

2. "Territorial Imperative," *Wall Street Journal*, Apr. 25, 1978.

3. "Korean Jet Shadowed, Brought Down in 1978," *Los Angeles Times*, Sept. 2, 1983.

4. "Agony of Flight 902," *Washington Post*, Apr. 24, 1978.
5. "Pilot in the '78 Incident Recalls Experience," *New York Times*, Sept. 9, 1983.
6. "Jet Survivors Say 2 Died in MiG Attack," *Washington Post*, Apr. 23, 1978.
7. "Pilot in '78 Incident," *New York Times*.
8. Ambassador Jeane J. Kirkpatrick, text, U.S. Mission to the United Nations, statement in the Security Council, Sept. 12, 1983, p. 2.
9. "Brzezinski Remark Stirs Fear of Security Breach," *New York Times*, Apr. 26, 1978.
10. "Soviets-Moscow Wanted Plane Down," *Washington Post*, Apr. 26, 1978.
11. "Agony of Flight 902," *Washington Post*.
12. "Odyssey Ends but Mystery Doesn't," *Washington Post*, Apr. 25, 1978.
13. "Jet Survivors Say 2 Died," *Washington Post*.
14. "Korean Airlines Flight Seven," *Review of the News*, Sept. 14, 1983.
15. "Plane Incidents Contain Eerie Similarities," *Washington Post*, Sept. 4, 1983.
16. "South Korean Jetliner Down in Soviet Union," *Washington Post*, Apr. 21, 1978.
17. *Congressional Record*, House, Sept. 14, 1983, p. H6876.
18. Lawrence P. McDonald, "Lessons from the History of Conquest of Byzantium," *American Opinion*, Jan. 1981, p. 25.

Chapter 4

1. Robert Conquest, "Brutality and Deceit: So What's New?" *Washington Post*, Sept. 11, 1983.
2. *Congressional Record*, Senate, Sept. 15, 1983, p. S12315.
3. Alexis de Tocqueville, *Democracy in America*, vol. 1, trans. Henry Reeve (New Rochelle, N.Y.: Arlington House), p. 431.
4. *Custine's Eternal Russia*, new edition of *Journey For Our Time*, with introduction by Ambassador Foy D. Kohler, ed. and trans. Phyllis Penn Kohler, Center for Advanced International Studies, University of Miami, Fla., 1976, Preface, p. 4.
5. Ibid., pp. 47, 93, 145.
6. Ibid., p. 60.
7. Ibid., pp. 209–211.
8. Ibid., Introduction, p. 9.
9. Robert Conquest, *The Great Terror, Stalin's Purge of the Thirties* (New York: Macmillan, 1968), pp. xiii–xiv.
10. Conquest, "Brutality and Deceit."
11. Dr. Richard Clayberg, interview with author, Oct. 6, 1983, Washington, D.C.
12. Lawrence McDonald, interview with author on "Reporter's Roundup," Mutual Broadcasting System, recorded in House Radio-TV Gallery, U.S. Capitol, Washington, D.C., July 21, 1977; Broadcast Mutual Radio Network, Aug. 7, 1977.
13. John Barron, *KGB, The Secret Work of Soviet Secret Agents* (New York: Bantam Books, 1974), pp. 450–51.
14. *Congressional Record*, extension of remarks, Mar. 14, 1983, p. E1018.

Chapter 5

1. Vladimir Solovyov and Elena Klepikova, *Yuri Andropov, A Secret Passage into the Kremlin* (New York: Macmillan, 1983), p. 210.

2. Kasim Gulek, author's radio interview, WRC Radio, Washington, D.C., May 15, 1981.
3. Martin Ebon, *The Andropov File, The Life and Ideas of Yuri Andropov, General Secretary of the Communist Party of the USSR* (New York: McGraw-Hill, 1983), p. 137.
4. Claire Sterling, "The Plot to Murder the Pope," *Reader's Digest,* Sept. 1982, pp. 72, 82.
5. Marvin Kalb, AP wire story, no. 51, New York, Sept. 14, 1982, 12:45 EDT.
6. "Pope's Threat to Quit over Poland Reported," *Washington Times,* Sept. 15, 1982.
7. "NBC News Links Soviets to Attempt on Pope's Life," *New York Times,* Sept. 15, 1982.
8. Solovyov & Klepikova, *Yuri Andropov,* pp. 209–10.
9. Alex Alexiev, "The Battle Between the Kremlin and the Pope," *Wall Street Journal,* Mar. 30, 1983.
10. Sen. Al D'Amato, TV interview by Jeffrey St. John and John Lofton, *Review of the News,* Washington, D.C., Feb. 23, 1983.
11. "CIA Inept on Pope Plot, D'Amato Says," *New York Times,* Feb. 8, 1983. Also AP dispatch, New York, Feb. 10, 1983.
12. Ebon, *Andropov File,* p. 138.
13. *Review of the News,* "A Capital Report," Feb. 23, 1983.
14. "Reactions to the Papal Shooting," *Wall Street Journal,* Feb. 2, 1983.
15. "Tass and the Pope" editorial, *Wall Street Journal,* Jan. 5, 1983.
16. Ibid.
17. Barron, *Secret Work,* p. 97.
18. Solovyov and Klepikova, *Yuri Andropov,* p. 210.
19. *Congressional Record,* House, "Andropov—The KGB Administration," Feb. 1, 1983, p. H228.

Chapter 6

1. "The Lonely Voice of Alexander Solzhenitsyn," *Wall Street Journal,* June 21, 1983.
2. Fred Smith, administrative assistant, Congressman Lawrence P. McDonald (D–Ga.), interview with author, Cannon House Office Building, U.S. Capitol, Washington, D.C., Sept. 23, 1983.
3. James Reston, "Reagan and Andropov," *New York Times,* Jan. 8, 1983.
4. Joseph Kraft, "Andropov Clearly Wants a Deal," *Washington Post,* Jan. 9, 1983.
5. "Soviet Encoding of Missile Data Assailed, Longstanding Dispute Revived," *Washington Post,* Jan. 6, 1983.
6. "Soviets Test New Missile, Possibly Violating SALT Terms," *Washington Post,* Feb. 16, 1983; and "Soviets Exceed SALT II Limits, U.S. Analysts Say," *Baltimore Sun,* Mar. 25, 1983.
7. "K.G.B. Officers Try to Infiltrate Antiwar Groups," *New York Times,* July 20, 1983.
8. "Defector Says Andropov Revamped KGB," *Washington Times,* U.P.I. story, Feb. 10, 1983.
9. Malcolm Toon, "What If the Charges Are True," *Washington Post,* Jan. 14, 1983.
10. "Percy Bids Reagan Meet Andropov Soon," *Baltimore Sun,* Mar. 16, 1983.

11. "Soviets Say Reagan Has 'Pathological Hatred,'" *New York Times*, Mar. 10, 1983.
12. "Reagan Accuses Soviets of Using Military to Extend Influence," *Washington Post*, Mar. 10, 1983.
13. "Panel Tells Reagan the Russians Seem to Have Broken Arms Pacts," *New York Times*, Apr. 21, 1983.
14. "Pen Pal Andropov Reassures U.S. Girl," *Baltimore Sun*, Apr. 26, 1983.
15. "Reagan Predicts Better Soviet Ties in Parley's Wake," *New York Times*, June 1, 1983.
16. "U.S. Discussing Arms Violations with Russians," *Washington Times*, May 20, 1983.
17. "Loose Talk about Soviet Cheating," *New York Times*, Apr. 25, 1983.
18. James Reston, "Moscow and the Summit," *New York Times*, June 1, 1983.
19. "Aides Urge Shultz to Visit Moscow to Test Its Mood," *New York Times*, June 12, 1983.
20. "Two Readings of the Climate Governing Relations between the Superpowers," *New York Times*, June 17, 1983.
21. "U.S.-Soviet Summit Is Doubtful, Gromyko Blames U.S. for Current Impasse," *Washington Post*, June 22, 1983.
22. "U.S.-Soviets Show Signs of Narrowing Rifts," *Washington Post*, June 18, 1983.
23. Edward Luttwak, "Delusion of Summitry," *New York Times*, July 27, 1983.
24. Tom Wicker, "Seizing the Initiative," "In the Nation" column, *New York Times*, Aug. 15, 1983.
25. "U.S. Lets Soviets Buy Pipelayers, Retains Grip on High-Tech Gear," *Washington Post*, Aug. 21, 1983.
26. Rowland Evans and Robert Novak, syndicated column by Field Enterprises, Inc., "The Push For a Summit," *Washington Post*, Aug. 24, 1983.
27. "Grain Pact Signed: U.S. Assures Soviet of Steady Supply," *New York Times*, Aug. 26, 1983.
28. "Soviet Leader Sees Reagan Meeting As Meaningless," *Washington Post*, Aug. 26, 1983.
29. "Lonely Voice of Solzhenitsyn," *Wall Street Journal*, June 21, 1983.
30. *Congressional Record*, extension of remarks, "Andropov Setting Stage for Move into Iran?" June 7, 1983, p. E-2747.

Chapter 7

1. Seweryn Bialer, "The Soviets Really Need Their Nukes," *Washington Post*, May 9, 1983.
2. Ted Agres, "Recon Plane Thwarted Soviet Missile Test—Off Course Flight 007 Flew over Planned Landing Zone," *Washington Times*, Sept. 12, 1983.
3. "Soviets Held Test of New Missile Three Days after Jet Downed," *Washington Post*, Sept. 16, 1983.
4. Frank Greve, "Soviet Site Hid Secret Missile, ABM Could Tilt Nuclear Balance," Knight-Ridder Newspapers, *Atlanta Journal and Constitution*, Sept. 11, 1983.
5. *Congressional Record*, Senate, Sept. 15, 1983, p. S12340.

6. Quoted in the *Congressional Record*, Senate, Mar. 23, 1983, pp. S3792–93.
7. "Soviets Exceed SALT II Limits, U.S. Analysts Say," *Baltimore Sun*, May 25, 1983.
8. *Congressional Record*, Senate, Mar. 23, 1983, p. S3789.
9. Rowland Evans and Robert Novak, "Reds Caught Cheating on SALT Again," *New York Post*, June 29, 1983.
10. "Executive Memorandum, Soviet SALT Cheating: The New Evidence," The Heritage Foundation, Washington, D.C., Aug. 5, 1983, no. 31.
11. Seymour Weiss, "But Let's Not Overlook the Hurdles," *Wall Street Journal*, Apr. 8, 1983.
12. Kenneth L. Adelman, Arms Control and Disarmament Agency director, interview by John Lofton, "Q&A, Reagan Administration's Goals in Arms Control Talks Outlined," *Washington Times*, Sept. 23, 1983.
13. "Secret Report Claims Carterites Hid Soviet Tricks to Aid SALT-2," *Washington Times*, Aug. 24, 1983.
14. "Moscow Offers Missile Pledge to Revive Talks," *Baltimore Sun*, datelined Moscow, Aug. 27, 1983.
15. "Soviets Doubling SS-20s Aimed at Near, Far East," *Washington Times*, May 10, 1983.
16. Ambassador Paul Nitze, press briefing transcript, the White House, Office of the Press Secretary, Sept. 3, 1983.
17. "Suspend the Arms Talks," editorial, *Wall Street Journal*, Sept. 13, 1983.
18. *Congressional Record*, Senate, Sept. 15, 1983, pp. S12285–86.
19. "U.S. Said to Weigh Arms Concession," *New York Times*, Sept. 21, 1983.
20. "Reagan Says U.S. Is Willing to Cut Back on Warheads It Wants to Base in Europe," *New York Times*, Sept. 27, 1983.
21. "Gromyko Accuses West of Blocking a Pact on Missiles," *New York Times*, Sept. 28, 1983.
22. "Andropov Blasts U.S. Military 'Militarist Course,'" *Washington Post*, Sept. 29, 1983.
23. Rowland Evans and Robert Novak column, "The President's SALT Screwdriver," *Washington Post*, Sept. 26, 1983.
24. Congressman Lawrence P. McDonald, interview with author, June 2, 1982, Cannon Office Building, U.S. Capitol, Washington, D.C.
25. Bialer, "Soviets Really Need Nukes."

Chapter 8

1. Sen. Jake Garn (R–Utah), *Korea in the World Today*, ed. Roger Pearson, Ph.D., "Conclusions and Recommendations for the Future," Council on American Affairs, Washington, D.C., 1976, pp. 87–88.
2. "Soviets Place Latest MiGs Near Japan," *Washington Times*, Aug. 31, 1983.
3. "Japan Sharply Protests Soviet Bullying," Text and unofficial translation, reprinted in *Wall Street Journal*, Jan. 31, 1983.
4. Ibid.
5. Rowland Evans and Robert Novak column, quoted in "Japan and Russia in War of Nerves," datelined Tokyo, *New York Post*, Sept. 9, 1983.
6. Robert Keatley, "Clumsy Soviet Diplomacy Toward China and Japan," *Wall Street Journal*, Feb. 7, 1983.
7. "Soviets Reported Preparing Rise in Asia Missiles," *New York Times*, May 8, 1983.

8. "Soviet Is Making More Use of Cam Ranh Bay," *New York Times*, Mar. 13, 1983.

9. "Why Soviets Are Sensitive about Northern Pacific Coast," *Christian Science Monitor,* Sept. 6, 1983.

10. Albert L. Weeks, "Soviet Navy Puts World within Kremlin Gasp," *New York Tribune*, Oct. 5, 1983.

11. Garn, *Korea in the World Today.*

12. Evans and Novak column, *New York Post*, Sept. 9, 1983.

13. "The Latest Victims," *Wall Street Journal*, Sept. 2, 1983.

14. The Hon. Lawrence McDonald, M.C., "The Republic of Korea: Frontline Ally of the Free World," transcript of undelivered speech, drafted Aug. 26, 1983, pp. 1–3.

15. "U.S. Urges Seoul to Exercise Restraint on Burma Bombing," *New York Times*, Oct. 13, 1983.

Chapter 9

1. "Media Conferee Faults U.S. Response on Jetliner," *New York Tribune*, datelined Cartagena, Colombia, Sept. 5, 1983.

2. Quoted in "Media Consensus on the Soviets," a column by R. Emmett Tyrrell, Jr., *Washington Post*, Sept. 12, 1983.

3. William C. Green and David B. Rivkin, Jr., paper titled "KAL 007 and Soviet Policy Priorities," Leon Sloss Associates, National Security Consultants, Arlington, Va., Sept. 14, 1983, p. 2.

4. Drew Middleton, "Strategic Soviet Region, Area Where Russians Say Plane Intruded Is Critical Part of Their Far East Defenses," *New York Times*, Sept. 2, 1983.

5. Albert L. Weeks, "Close Analysis Shows Order to Destroy Jet Came from Top Level," *New York Tribune*, Sept. 6, 1983.

6. Ibid.

7. John Barron, *KGB, The Secret Work of Soviet Secret Agents* (New York: Bantam, 1974), p. 110.

8. "Soviets Say Nixon Had Been Booked on Flight 007," AP Moscow dispatch, *Washington Post*, Sept. 25, 1983.

9. Garrett N. Scalera, interview with author, Washington, D.C., Oct. 7, 1983.

10. Ibid.

11. "U.S. Intelligence Men Suggest Soviets Downed the Wrong KAL 747," *New York Tribune*, Sept. 2, 1983.

12. Kathryn McDonald, interview with author, Hot Springs, Va., Sept. 9, 1983.

13. "Q&A: Kathryn McDonald: The Commitments Continue," *Washington Times*, Sept. 8, 1983.

14. Bruce Herbert, "The Real Mystery of KAL 007," Center for International Security paper, Sept. 20, 1983, p. 5.

15. Bruce Herbert, interview with author, Center for International Security, Washington, D.C., Sept. 23, 1983.

16. Herbert, "Real Mystery of KAL 007."

17. Herbert, interview with author.

18. "Explaining the Inexplicable," *Time*, Sept. 19, 1983.

19. Dr. Igor Glagolev, interview with author, Oct. 20, 1983, Washington, D.C.

20. Robert Reilly, "A Terrorist Act with a Political Purpose," commentary section, *Washington Times*, Sept. 8, 1983.
21. "Media Conferee Faults U.S. Response on Jetliner," *New York Tribune*.
22. Lawrence McDonald, "A Talk with Major General George Keegan: On Defense," *American Opinion*, Sept. 1977, p. 79.

Chapter 10

1. Sen. Steve Symms (R–Ida.), interview with author, Oct. 3, 1983, Washington, D.C.
2. J. Lynn Helms, FAA administrator, full text of statement before the Extraordinary Session of the Council of International Civil Aviation Organization (ICAO), Sept. 15, 1983, Montreal, Canada.
3. *Korea Newsreview*, Oct. 1, 1983, p. 18.
4. *Korea Newsreview*, Sept. 10, 1983, p. 5.
5. *Massacre in the Sky—The Soviet Downing of a KAL Passenger Plane*, "The List of Passengers and Crew" (names and ages), Korean Overseas Information Service, Seoul, Korea, Sept. 1983, pp. 72–76.
6. "Incomprehensible Noise, Then Awful Silence," *U.S. News & World Report*, Sept. 12, 1983.
7. Lawrence McDonald, text of "The Republic of Korea: Frontline Ally of the Free World," drafted Aug. 26, 1983, p. 6.
8. "North Korean Spy Ship Report Sunk by S. Korea," *New York Tribune*, Aug. 15, 1983.
9. "Metro KAL Plane Toll Is 33," *New York Tribune*, Sept. 3, 1983.
10. *Congressional Record*, House, Sept. 14, 1983, p. H6890.
11. Ibid., p. H6873.
12. R. D. Patrick Mahoney, interview with author, Cannon House Office Building, U.S. Capitol, Washington, D.C., Sept. 23, 1983.
13. Congressman Carroll Hubbard, interview with author, Rayburn House Office Building, U.S. Capitol, Washington, D.C., Oct. 4, 1983.
14. Sen. Steven Symms and Frances E. Symms, interview with author, Washington, D.C., Oct. 3, 1983.
15. Ibid.
16. Joseph C. Morecraft III, "The Meaning of the Life and Death of Congressman Lawrence Patton McDonald," Multi-Media Ministries, Atlanta, Ga., 1983, p. 5.

Chapter 11

1. Jack Anderson, "Materials in Files Add to Mystery of Jet Downing," *Washington Post*, Sept. 20, 1983.
2. "The Agony of Flight 902," *Washington Post*, Apr. 24, 1978.
3. "Korean Airlines Flight Seven," *Review of the News*, Sept. 14, 1983.
4. Reprint of Federal Aviation navigation map, *Aviation Week & Space Technology*, Sept. 12, 1983, p. 19.
5. "FAA Studies Upgraded Pacific Nav-aids," *Aviation Week & Space Technology*, Sept. 19, 1983, p. 18.
6. "How Flight 007 Strayed," *Newsweek*, Sept. 19, 1983, p. 29.
7. "Korean Plane's Pilot Was Air Force Veteran," *New York Times*, Sept. 6, 1983.
8. "Fatal Error By Star Pilot? Korean Airlines Man Was One of Best in Sys-

tem," *New York Daily News*, Sept. 14, 1983. Also: "Korean Plane's Pilot,"*New York Times*.

9. "Korean Plane's Pilot," *New York Times*.

10. Lt. Gen. Daniel O. Graham, U.S.Army (Ret.), interview with author, Washington, D.C., Sept. 19, 1983.

11. Col. Samuel T. Dickens, U.S.A.F. (Ret.), interview with author, Washington, D.C., Sept. 29, 1983.

12. "Guidance System Seen Fail Safe," *Washington Times*, Sept. 2, 1983.

13. AP dispatch, Washington, D.C., reprinted in the *Korea Herald* (Seoul), Sept. 6, 1983.

14. *Aviation Week & Space Technology*, pp. 18–20.

15. Anderson, "Materials in Files."

16. "Human Factors Analyzed in 007 Navigation Error," *Aviation Week & Space Technology*, Oct. 3, 1983, p. 165.

17. William Hendrick, interview with author, Oct. 26, 1983, Washington, D.C.

18. "How 007 Strayed," *Newsweek*.

19. Anderson, "Materials in Files."

20. "Jetliner Forced Down On Soviet-Held Island, Rep. McDonald Believed Aboard," dispatch, from Seoul, Korea, published in *New York Tribune*, Sept. 1, 1983.

21. "Korean Plane's Pilot," *New York Times*.

22. Sen. Steve Symms and Frances Symms, interview with author, Washington, D.C., Oct. 3, 1983.

Chapter 12

1. Col. Samuel T. Dickens, U.S.A.F. (Ret.), interview with author, Sept. 29, 1983, Washington, D.C.

2. "Soviet Downing of a South Korean Airliner; Background and Issues," Congressional Research Service, Library of Congress, Washington, D.C., Sept. 21, 1983. Chronology of Events, p. CRS-18.

3. J. Lynn Helms, FAA administrator, text of statement before the Extraordinary Session of the Council of the International Civil Aviation Organization (ICAO), Sept. 15, 1983, Montreal, Canada, p. 5.

4. Ibid., p. 7.

5. "U.S. Intercepts of Soviet Fighter Transmissions," reprinted in *Aviation Week & Space Technology*, Sept. 12, 1983, pp. 22–23.

6. "Pilot Who Shot Down Korean Jet Is Interviewed On Soviet Television," *New York Times*, Sept. 11, 1983.

7. "Editor's Note on Text of Transcript," *New York Tribune*, Sept. 7, 1983.

8. "Soviet Envoy Calls Tapes 'Fabrication,'" Reuter's dispatch, *Korea Herald*, U.S. ed., New York, Sept. 10, 1983.

9. "Spellbinding Performance in Moscow," *New York Times*, Sept. 10, 1983.

10. "The Army's 'Main Brain,'" *Newsweek*, Sept. 19, 1983.

11. Viktor Suvorov, *Inside the Soviet Army* (New York: Macmillan, 1982), p. 104.

12. "The Downed Plane, An Interview with Bobby R. Inman, What Probably Happened," *Washington Post*, Sept. 4, 1983.

13. Col. Dickens, interview with author.

14. "Soviet Pilot Knew Target Was Jetliner, Sources Say, Intelligence Intercepts Called 'Hard Evidence,'" *Washington Times*, Sept. 29, 1983.
15. Col. Dickens, interview with author.
16. Ibid.
17. "Jetliner Spiraled for 12 Minutes, 2nd Japanese Radar Watched Airliner Crash," AP dispatch, *Korea Herald*, U.S. ed., New York, Sept. 16, 1983.
18. "Last Minutes in the Killing of KAL Flight 007," *Washington Times*, Sept. 5, 1983.
19. "Tail Part of KAL Plane Found, Body of Child Also Recovered Off Hokkaido," AP dispatch, *Korea Herald*, U.S. ed., New York, Sept. 13, 1983.
20. Letter from the chairman, *FALN: Threat to America*, Aug. 1981, Western Goals Foundation, Alexandria, Va., 1981.

Chapter 13

1. Text of statement by Ambassador Jeane Kirkpatrick, U.S. Permanent Representative to the United Nations in the U.N. Security Council, Sept. 6, 1983, p. 12.
2. Quoted in "Island a Focus of Russia-Japan Disputes," *New York Times*, Sept. 2, 1983.
3. Aleksandr I. Solzhenitsyn, *The Gulag Archipelago, Three* (New York: Harper and Row, 1976), p. 10.
4. "The Forsaken People: Koreans Held in Sakhalin," *New York Times*, Oct. 3, 1983.
5. Solzhenitsyn, *Gulag Archipelago*, p. 525.
6. Helms, text of statement before ICAO, p. 10.
7. "Abe Laments Russ Preying on Airliner," UPI dispatch, Tokyo, *Korea Herald*, U.S. ed., New York, Sept. 4, 1983.
8. "Bits of Wreckage of the Airliner Recovered," UPI dispatch, Tokyo, *Korea Herald*, U.S. ed., New York, Sept. 6, 1983.
9. "Search Continuing for Traces of KAL Victims," *Korea Herald*, U.S. ed., New York, Sept. 8, 1983.
10. "Tail Part of KAL Plane Found," Wakkanai, Japan, AP dispatch, *Korea Herald*, Sept. 13, 1983.
11. "Black Box Recovery Soon, U.S. Informs Seoul of KAL Salvage Progress," *Korea Herald*, Sept. 24, 1983.
12. "Soviets Turn Over Bits of Jet Flotsam," *Washington Times*, Sept. 27, 1983.
13. "Soviet Hands Over Debris from Korean Plane," *New York Times*, Sept. 27, 1983.
14. "KAL Victim Families Hold Chusok Rite Near Crash Site," UPI dispatch aboard the *Soya Maru* No. 5, *Korea Herald*, Sept. 24, 1983.
15. "Families Mourn Loved Ones on Sea," *Korea Newsreview*, Seoul, Korea, Sept. 10, 1983, pp. 14–15.
16. "U.S. Ships Pick Up Black Box Beeps, Soviets Obstruct KAL Wreckage Search Operations: Pentagon," AP dispatch, *Korea Herald*, Sept. 23, 1983.
17. "U.S. Admiral Optimistic about Search Results," AP dispatch, Wakkanai, Japan, *Korea Herald*, Sept. 24, 1983.
18. "U.S. Says Soviet Ships Harass Plane-Data Searchers," *Washington Post*, Sept. 21, 1983; and "Navy Says Search Hampered by Soviets," *Washington Times*, Sept. 21, 1983.

19. "Soviets Threaten Searchers, Japanese Ship Aided By U.S.," *Washington Post* dispatch, *Dallas Times Herald*, Oct. 13, 1983.

20. "Soviets Send Bombers toward Japan Island," UPI dispatch, *Washington Times*, Sept. 14, 1983.

21. Kirkpatrick, text of statement, p. 12.

22. "By 26–2 Vote, ICAO Deplores Soviets' Shooting of Plane," Montreal, *Washington Post*, Sept. 17, 1983.

23. "ICAO Resolution Calls for Full Probe of KAL Downing," *Korea Newsreview*, Seoul, Korea, Sept. 24, 1983.

24. "Soviet Refuses Compensation Again," *Korea Newsreview*, Seoul, Korea, Sept. 24, 1983.

25. "Kennedy Interviews Congressman Lawrence P. McDonald," *Western Monetary Report*, Aug. 15, 1983, reprinted in full-page advertisement titled "America Is *Still* Outraged!" The Conservative National Committee, Fort Collins, Colo., *Washington Times*, Sept. 23, 1983.

Chapter 14

1. President Chun Doo Hwan, text of special statement at the memorial service for the victims of the downed KAL plane, Embassy of Korea, Washington, D.C., Sept. 7, 1983, p. 5.

2. "Korea Ambassador: 'I Knew, I Had a Fear,'" *New York Times*, Sept. 8, 1983.

3. Ibid.

4. Sen. Steven Symms, interview with author, Oct. 3, 1983, Washington, D.C.

5. Frances E. Symms, interview with author, Oct. 3, 1983, Washington, D.C.

6. Steven Symms, interview with author.

7. Sen. Orrin Hatch (R–Utah), text of speech, "What Should the Friends of South Korea Be Doing?" Shulla Hotel, Seoul, Korea, Sept. 3, 1983, p. 2.

8. Sen. Orrin Hatch, interview with author, Sept. 26, 1983, Washington, D.C.

9. "Tokyo Is Carefully Angry—Soviet Lack of Sincerity," *New York Times*, Sept. 18, 1983.

10. "Korea Ambassador: 'I Knew, I Had a Fear,'" *New York Times*.

11. Text, address of the president to the nation, the Oval Office, Sept. 5, 1983, 8:00 P.M. EDT, the White House, Office of the Press Secretary, Washington, D.C., p. 7.

12. President Chun Doo Hwan, text of special statement.

13. "Protests Sweep World Over Soviet Downing of KAL Jet," *Korea Newsreview*, Sept. 10, 1983, p. 5.

14. "Government Requests Soviet Compensation for KAL Jet," *Korea Newsreview*, Sept. 17, 1983, p. 7.

15. "Catholic Priest Writes Song Condemning Soviet Attack on KAL," *Korea Newsreview*, Sept. 24, 1983, p. 26.

16. "Song Kwang-sun to Sing Song Dedicated to KAL Victims," *Korea Herald*, Oct. 6, 1983.

17. Stephen Cardinal Kim Sou-hwan, "KAL Incident Shouldn't Zap Zeal for Humanity: Cardinal," English translation of article appearing in Dong-A Ilbo newspaper, *Korea Herald*, Sept. 23, 1983.

18. "Burma Says Agents of North Korea Set Blast That Killed 21," A.P., Rangoon, Burma, *New York Times*, Nov. 5, 1983.
19. "H.K. Girl Asks Andropov to Explain Friend's Death," UPI dispatch, Hong Kong, *Korea Herald*, Sept. 8, 1983.
20. Steven Symms, interview with author.
21. Lawrence P. McDonald, "The Coming Anti-Communist Alliance," *American Opinion*, Mar. 1978, p. 87.

Chapter 15

1. Sen. Orrin Hatch (R–Utah), interview with author, Sept. 26, 1983, Washington, D.C.
2. "Soviets Confident Furor Will Pass," *New York Times*, Sept. 12, 1983.
3. George P. Shultz, Secretary of State, "On Downing of Korean Commercial Aircraft," Department of State Release No. 327, press briefing, U.S. State Department, Washington, D.C., Sept. 1, 1983, 10:45 A.M., transcript, p. 2.
4. Ibid., p. 4.
5. "Reagan Was Out of Touch with Situation All Night," *Baltimore Sun*, Sept. 2, 1983.
6. "Transcript of Reagan's Statement on Airliner," Sept. 2, 1983, Point Mugu, California, *New York Times*, Sept. 3, 1983.
7. "Mrs. McDonald Asks Reagan to Speak at Memorial Service," *Atlanta Journal and Constitution*, Sept. 4, 1983.
8. Text, address of the president to the nation, the Oval Office, Sept. 5, 1983, 8:00 P.M. EDT, the White House, Office of the Press Secretary, p. 1.
9. William Safire, "Sticks and Stones," *New York Times*, Sept. 8, 1983.
10. "An Interview with President Reagan," *Time*, Sept. 19, 1983.
11. William H Gregory, "Fear of Defections," editorial, *Aviation Week & Space Technology*, Sept. 12, 1983.
12. Congressman Newt Gingrich (R–Ga.), interview with author, Oct. 18, 1983, Washington, D.C.
13. Hatch, interview with author.
14. "Inquest on a Massacre," *Newsweek*, Sept. 19, 1983.
15. "Reagan Vs. the New Right," *Newsweek*, Sept. 19, 1983.
16. Ibid.
17. "Reagan Rebukes His KAL Reaction Critics," *New York Tribune*, Sept. 10, 1983.
18. Howard Phillips, text of remarks prepared for delivery at McDonald Memorial Service, Constitution Hall, Washington, D.C., Sept. 11, 1983, p. 4.
19. Gingrich, interview with author.

Chapter 16

1. John Rees, "Memorial for a Warrior," *Review of the News*, Sept. 21, 1983, pp. 45–46.
2. "Reagan Sets National Day of Mourning," *New York Tribune*, Sept. 10, 1983.
3. "Reagan Declares Day of Mourning," *New York Times*, Sept. 10, 1983.
4. George Will, ". . . For Good Reason," *Washington Post*, Sept. 18, 1983.
5. "Soviets Confident Furor Will Pass," *New York Times*, Sept. 12, 1983.
6. "Inquest on a Massacre," *Newsweek*, Sept. 19, 1983.

7. "Conservatives Pay Homage to Larry McDonald," *Human Events,* Sept. 24, 1983.
8. Rees, "Memorial," p. 40.
9. "Actions Which Ought to Be Taken by U.S. Government," news release from the Conservative Caucus, Inc., Vienna, Va., Sept. 11, 1983.
10. "Nation Is Confused about Jet Downing, Latest Poll Suggests," *New York Times,* Sept. 16, 1983.
11. "A Newsweek Poll: Get Tougher with the Soviets," *Newsweek,* Sept. 19, 1983.
12. John Lofton, "Conservatives Do Have an Alternative," *Washington Times,* Sept. 21, 1983.
13. Congressman Newt Gingrich (R–Ga.), interview with author, Washington, D.C., Oct. 18, 1983.
14. Rees, "Memorial."
15. Otto J. Scott, interview with author, Sacramento, Calif., Oct. 15, 1983.

Chapter 17

1. Burton Pines, transcript of interview, ABC "Nightline," Sept. 20, 1983, ABC Network.
2. Press Release, U.S. Mission to the United Nations, "United States Requests Urgent Security Council Meeting," Sept. 1, 1983.
3. U.N. Security Council, Provisional Verbatim Record, U.N. Headquarters, Friday, Sept. 2, 1983, pp. 32–35.
4. Ibid., p. 31.
5. Ibid., p. 72.
6. Ibid., p. 73.
7. Ibid., p. 47.
8. Bernard D. Nossiter, "Moscow Vetoes U.N. Criticism in Plane Affair," *New York Times,* Sept. 13, 1983.
9. "U.S. Encounters Political Snag in Security Council," *New York Times,* Sept. 8, 1983.
10. "Text of Council Resolution on Downing of the Airliner," *New York Times,* Sept. 13, 1983.
11. Nossiter, "Moscow Vetoes U.N. Criticism in Plane Affair."
12. Ibid.
13. "Shultz Rejects Zimbabwe Aid Cuts, Country Abstained on Soviet Vote," *Washington Post,* Oct. 22, 1983.
14. "U.S. Considers Cuts in Aid To Zimbabwe, U.S. Vote Backlash," *Washington Post,* Oct. 13, 1983.
15. "9 U.N. Votes Seen as Success for U.S., Allies over Soviets," *New York Tribune,* Sept. 14, 1983.
16. "Kremlin Decides Gromyko Won't Go to U.N. Assembly," *New York Times,* Sept. 18, 1983.
17. "Gromyko's Move Casts Shadow at U.N.," *New York Times,* Sept. 18, 1983.
18. Ibid.
19. George F. Will, "Needed: A Policy of Punishment," *Newsweek,* Sept. 12, 1983.
20. "U.N. Aid Suggests Members Take U.N. Elsewhere If They So Desire," *New York Times,* Sept. 20, 1983.

21. "Support Rolling in to Kiss the U.N. Goodbye," *New York Post*, Sept. 21, 1983.
22. "US-USSR Relations Still the Same Despite 007 Incident, Says Kirkpatrick," *New York Tribune*, Sept. 23, 1983.
23. "Reagan Rejects Lichenstein's Resignation," *Washington Times*, Sept. 27, 1983.
24. "Reagan Raises Idea U.N. Spend Time in Moscow," *New York Times*, Sept. 22, 1983.
25. "Senate Votes to Cut U.N. Contribution," *Washington Post*, Sept. 23, 1983.
26. "U.N. Budget: High Pay, High Costs, Move to Cut U.N. Share," *Los Angeles Times*, Oct. 15, 1983.
27. "White House Rushes to Aid of U.N. after Funds-Cut Vote," *Washington Post*, Sept. 24, 1983.
28. "Text of President Reagan's Address to U.N. General Assembly," *Washington Post*, Sept. 27, 1983.
29. "U.S. Envoy to U.N. Criticizes Soviet," *New York Times*, Oct. 5, 1983.
30. Burton Pines, transcript of interview, ABC "Nightline."
31. "U.N.—Success or Failure?" *U.S. News & World Report*, interview with Congressman Larry P. McDonald, Sept. 17, 1979, pp. 71–72.

Chapter 18

1. *Congressional Record*, House, Sept. 14, 1983, p. H6885.
2. Ed Rogers, "U.S. Outraged by Murder of Passengers on KAL-007," *New York Tribune*, Washington Times News Service story, Sept. 2, 1983.
3. *Congressional Record*, House, p. H6864.
4. Ibid.
5. Ibid., p. H6866.
6. Ibid., p. H6868.
7. Ibid., p. H6874.
8. Ibid., p. H6885.
9. Ibid.
10. Congressman Larry McDonald, Seventh District, Georgia, "Rep. McDonald Criticizes Intolerance by Liberals," Information-News Report, newsletter released Jan. 7, 1983, p. 2.
11. Member of the U.S. House of Representatives, interview with author, Oct. 4, 1983.
12. McDonald, "Criticizes Intolerance by Liberals."
13. *Congressional Record*, House, p. H6883.
14. Ibid., p. H6885.

Chapter 19

1. *Congressional Record*, Senate, Sept. 15, 1983, p. S12305.
2. Roger Parkinson, *Peace for Our Times* (New York: David McKay, 1971), p. 66. Parkinson quotes President Roosevelt from a cable sent to Neville Chamberlain after he and Hitler had signed the Munich agreement that contained two words: "Good man."
3. Telford Taylor, *Munich, The Price of Peace, The Definitive Account of the Fateful Conference of 1938* (Garden City, New York: Doubleday, 1979), pp. xiv–xv.

4. *Congressional Record*, Senate, Sept. 15, 1983, p. S12304.
5. Ibid., p. S12313.
6. Ibid.
7. Ibid., p. S12314.
8. Ibid., p. S12261.
9. Ibid.
10. Ibid., p. 12312.
11. Ibid., p. S12274.
12. Ibid., p. S12321, roll call vote no. 245–yeas 70, nays 25, not voting 5. Roll call vote no. 246—yeas 69, nays 26, not voting 5.
13. Ibid., p. S12322, roll call vote no. 247—yeas 50, nays 45, not voting 5. Roll call vote no. 248—yeas 60, nays 36, not voting 4. Roll call vote no. 249—yeas 60, nays 36, not voting 4.
14. Ibid., pp. S12322–23, roll call vote no. 250—yeas 66, nays 29, not voting 5. Roll call vote no. 251—yeas 52, nays 43, not voting 5.
15. Ibid., p. S12327, roll call vote no. 252—yeas 49, nays 45, not voting 4.
16. Ibid., p. S12323.
17. Ibid., p. S12343, roll call vote no. 253—yeas 95, nays 0, not voting 5.
18. Ibid., pp. S12280–81.
19. Ibid., p. S12278.
20. Ibid., pp. S12338–39.
21. Ibid., p. S12265.
22. Hedrick Smith, *The Russians* (New York: Ballantine, 1976), p. 335.
23. Taylor, *Munich*, pp. 10003–04.

Chapter 20

1. Quoted in "Administration Blocks Attempt to Ban Soviet Imports," *Washington Times*, Oct. 27, 1983.
2. Statement by the president, the White House, Office of the Press Secretary, for immediate release, Sept. 15, 1983, p. 1.
3. "Wings Clipped, Aeroflot Closes Shop," *Washington Post*, Sept. 16, 1983.
4. "Canada Bars Aeroflot Landings," *Washington Times*, Sept. 6, 1983.
5. William H. Gregory, "Outrage from the Pilots," an editorial, *Aviation Week & Space Technology*, Sept. 19, 1983.
6. Michael Dobbs, "Canada Suspends Soviet Flights; Paris Postpones Gromyko Visit," *Washington Post*, Sept. 6, 1983.
7. "European Pilots Join Ban on Moscow," *New York Times*, Sept. 8, 1983.
8. "Pilots Begin Soviet Boycott; Bonn, Madrid, Tokyo Set Ban," *Washington Post*, Sept. 13, 1983.
9. "Pilots Urge End to the Boycott of Soviet Flights, Willingness of Moscow to Cooperate Is Cited," *New York Times*, Oct. 1, 1983.
10. "European Pilots Join Ban," *New York Times*.
11. "Aeroflot Nets Gains Despite U.S. Sanctions," *Washington Times*, Sept. 26, 1983.
12. "Denial Urged on Oil Fear for Soviet," *New York Times*, Sept. 20, 1983.
13. "State Ordered to Review Controls on Strategic Items," *Washington Post*, Sept. 23, 1983.
14. William Safire, "Selling the Rope," *New York Times*, Oct. 9, 1983.
15. Ibid.
16. Dr. Miles Costick, interview with author, Nov. 16, 1983, Washington, D.C.

17. "Curb Asked on Trade to Soviet, Clark Supports Stricter Rules," *New York Times*, Sept. 22, 1983.
18. Rowland Evans and Robert Novak, "High Cost of Shifting Clark," column syndicated by Field Enterprises, Inc., *Washington Post*, Oct. 17, 1983.
19. "U.S. Asked to Ban Soviet Items Made by Forced Labor," *New York Times*, Oct. 8, 1983.
20. "Administration Blocks Attempt to Ban Soviet Imports," *Washington Times*, Oct. 27, 1983.
21. Ibid.
22. "Grain Sales to Soviet Increasing," *New York Times*, Sept. 19, 1983.
23. Victor Riesel, "Soviet Freezing U.S. Freighters Out of Grain Business," "Inside Labor" column, Field Newspaper Syndicate, Inc., Sept. 29, 1983.
24. Larry McDonald, "Technology and Our Enemies," Seventh District, Georgia, information news report, column for release June 14, 1983.

Chapter 21

1. Ronald Reagan television address, "A Time for Choosing," Oct. 27, 1964, reprinted in *Ronald Reagan Talks to America* (Greenwich, Conn.: Devin Adair, 1983), p. 17.
2. Quoted in *The Real Reagan*, Frank van der Linden (New York: William Morrow, 1981), p. 123.
3. Author's notes and telephone log, Nov. 1, 1983.
4. Text of answers of Secretary of State George P. Shultz regarding the KAL 007 tragedy given to Jeffrey St. John, dated Nov. 7, 1983, question and answer #3.
5. Ibid., question and answer #4.
6. Ibid., #2 and 5.
7. Ibid., #6.
8. Ibid., #7.
9. Ibid., #8.
10. Ibid., #12.
11. Ibid., #16.
12. Ibid., #18.
13. Ibid., #19.
14. Reagan, "A Time for Choosing," p. 17.
15. Acceptance speech by Governor Ronald Reagan, Republican National Convention, Detroit, Michigan, July 17, 1980, reprinted in *Ronald Reagan, A Political Biography*, Lee Edwards (Houston: Nordland Publishing International, 1981), p. 283.
16. Inaugural address of President Ronald Reagan, Jan. 20, 1981, reprinted in *Ronald Reagan*, Edwards, p. 291.
17. Steven R. Weisman, "Reagan Campaign Advisors Say He Will Face a Tough Race in '84," *New York Times*, Sept. 17, 1983.
18. Morton Kaplan, "Papering Over as Foreign Policy," *Washington Times*, Nov. 17, 1983.
19. Lawrence McDonald, "Central America," *American Opinion*, June 1983, pp. 1–2, 109.

Chapter 22

1. Jeane Kirkpatrick, "The Mass Media: Can It Handle Democracy?" interview with journalist-historian George Urban, *New York Post*, Oct. 25, 1983.

2. George Orwell, "The Prevention of Literature," Jan. 1946, reprinted in *The Collected Essays, Journalism and Letters of George Orwell*, vol. 4, *In Front Of Your Nose, 1945–1950* (New York: Harcourt, Brace, Jovanovich, 1968), p. 64.
3. "The Downed Airliner: Why?" editorial, *Washington Post*, Sept. 2, 1983.
4. "Murder in the Air," editorial, *New York Times*, Sept. 2, 1983.
5. Meg Greenfield, "Our Square-One Complex," *Newsweek*, Sept. 19, 1983.
6. "From Lies to Remedies," editorial, *New York Times*, Sept. 7, 1983.
7. Otto J. Scott, interview with author, Sacramento, Calif., Oct. 14, 1983.
8. Col. Samuel Dickens, USAF (Ret.), interview with author, Washington, D.C., Sept. 29, 1983.
9. Ellen Goodman, column syndicated by Boston Globe Newspaper Company, "Flight 007: Not a Call to Arms," *Washington Post*, Sept. 17, 1983.
10. Orwell, "Lear, Tolstoy, and the Foot," Mar. 1947, reprinted in the *Collected Essays*, p. 301.
11. R. Emmett Tyrrell, Jr., "The Media Consensus on the Soviets," *Washington Post*, Jan. 12, 1983.
12. Kirkpatrick, "The Mass Media."
13. Congressman Lawrence P. McDonald, interview with author, Cannon Office Building, U.S. Capitol, Washington, D.C., June 2, 1983.
14. Orwell, "Letter to Francis A. Henson," June 16, 1949, reprinted in *Collected Essays*, p. 502.

Chapter 23

1. Mrs. Kathryn J. McDonald, interview with author, Alexandria, Va., Nov. 15, 1983.
2. "U.S. Ends Search for KAL Jet," *Washington Times*, Nov. 7, 1983.
3. Kathryn McDonald, interview with author.
4. Ibid.
5. Dr. Harold McDonald, interview with author, Atlanta, Ga., Nov. 18, 1983.
6. *The Political Report*, vol. 6, no. 45, Free Congress Research and Education Foundation, Nov. 11, 1983.
7. "McDonald to Face Darden in November 8 Runoff," *Congressional Quarterly*, Washington, D.C., Oct. 22, 1983, p. 2185.
8. Lawrence McDonald, full text of testimony before the Committee on Foreign Relations, Washington, D.C., Jan. 25, 1977. Reprinted in the *Congressional Record*, House, Feb. 2, 1977.
9. Senator Robert Byrd, quoted in AP wire dispatch by H. Josef Herbert, Washington, D.C., July 17, 1978.
10. "Impeach Andrew Young, Hon. Larry McDonald of Georgia in the House of Representatives," *Congressional Record*, July 13, 1978.
11. Henry Kissinger, quoted in AP wire dispatch by H. Josef Herbert, Washington, D.C., July 17, 1978.
12. "Ambassador Young's Offense," *New York Times*, editorial, Aug. 16, 1979.
13. Ambassador Andrew Young, quoted in "Young Back in Doghouse," *Atlanta Constitution*, Feb. 9, 1979.
14. Tommy Toles, interview with author, Rome, Ga., Nov. 25, 1983.
15. Ibid.
16. Ibid.
17. *Political Report*.

18. Kathryn McDonald, quoted in "The Widow and Her Fight for the Right; Kathy McDonald Takes Up the Torch after Flight 007," by Art Harris, *Washington Post,* Oct. 13, 1983.
19. Kathryn McDonald, interview with author.
20. Ibid.

Chapter 24

1. Congressman Lawrence McDonald, interview with author, Cannon Office Building, U.S. Capitol, Washington, D.C., June 2, 1982.
2. David Leon Chandler, *The Natural Superiority of Southern Politicians* (Garden City, New York: Doubleday, 1977), Introduction, p. 2.
3. Ibid., p. 368.
4. McDonald, interview with author.
5. McDonald, *We Hold These Truths* (Seal Beach, Calif.: '76 Press, 1976), p. 13.
6. Dr. Daniel Jordan, interview with author, Atlanta, Ga., Nov. 14, 1983.
7. Ibid.
8. Dick Williams, "McDonald—An Extremist in Defense of Liberty," *Atlanta Journal,* Sept. 11, 1983.
9. Rev. Joseph C. Morecraft III, interview with author.
10. "McDonald's War, Hate Made His World Go 'Round," *New Republic,* Oct. 3, 1983.
11. Bruce Herbert, interview with author, Washington, D.C., Sept. 23, 1983.
12. Kathryn McDonald, interview with author, Alexandria, Va., Nov. 15, 1983.
13. McDonald, interview with author.
14. Paul Weyrich, "McDonald's Death Must Galvanize Us," *Washington Times,* Sept. 5, 1983.

Epilogue

1. Tape recording of a speech by Congressman Lawrence P. McDonald, "A Time for Action," June 1983, Marietta, Ga.
2. Congressman Lawrence P. McDonald, interview with author, The Cannon Office Building, U.S. Capitol, Washington, D.C., June 2, 1982.
3. *The Federalist Papers* (New York: New American Library, 1961), "Federalist 10, Madison," p. 81.
4. Will and Ariel Durant, *The Lessons of History* (New York: Simon and Schuster, 1968), p. 100.
5. Congressman Newt Gingrich (R–Ga.), interview with author, Washington, D.C., Oct. 18, 1983.
6. Congressman Newt Gingrich with Dr. Steven Hanswer, Walter Jones, and David Warwick, *On Survival,* draft, unpublished manuscript, June 1983, pp. 144–45.
7. "U.S. Confusing Kremlin Leaders, Kissinger Says," *Washington Times,* Oct. 18, 1983.
8. Elmo R. Zumwalt, Jr., *On Watch* (New York: Quadrangle/New York Times Book Co., 1976), pp. xiv–xv.
9. "Carter and Ford Oppose U.S. Strike," *New York Times,* Nov. 7, 1983.
10. President Ronald R. Reagan, letter to the Honorable Meldrim Thomson, Jr., the White House, Washington, D.C., Oct. 3, 1983, p. 1.

11. Lawrence P. McDonald, "Central America," *American Opinion*, June 1983, p. 2.

12. Otto J. Scott, interview with author, Sacramento, Calif., Oct. 14, 1983.

13. Congressman Lawrence P. McDonald, text of a speech, "The Demise of U.S. Internal Security," American Conservative Union, Washington, D.C., July 11, 1978, p. 19.

14. Raymond Kline, interview, MacNeil-Lehrer News Hour, PBS, WETA–TV, Washington, D.C., Nov. 25, 1983.

15. "Reagan Security," Don Phillips, UPI dispatch, Washington, D.C., Jan. 25, 1984.

16. Text of Message from the president on the State of the Union, *New York Times*, Jan. 26, 1984.

17. McDonald, "A Time for Action."

18. Ibid.

19. Rev. Joseph Morecraft III, interview with author, Sacramento, Calif. Oct. 14, 1983.

20. Dr. Harold McDonald, Jr., interview with author, Atlanta, Ga., Nov. 18, 1983.